T0214518

Lecture Notes in Computer Science 11510

Commenced Publication in 1973
Founding and Former Series Editors:
Gerhard Goos, Juris Hartmanis, and Jan van Leeuwen

More information about this series at http://www.springer.com/series/7412

Donatello Conte · Jean-Yves Ramel ·
Pasquale Foggia (Eds.)

Graph-Based Representations in Pattern Recognition

12th IAPR-TC-15 International Workshop, GbRPR 2019
Tours, France, June 19–21, 2019
Proceedings

 Springer

Editors
Donatello Conte (iD)
University of Tours
Tours, France

Jean-Yves Ramel
University of Tours
Tours, France

Pasquale Foggia (iD)
University of Salerno
Fisicano, Italy

ISSN 0302-9743 ISSN 1611-3349 (electronic)
Lecture Notes in Computer Science
ISBN 978-3-030-20080-0 ISBN 978-3-030-20081-7 (eBook)
https://doi.org/10.1007/978-3-030-20081-7

LNCS Sublibrary: SL6 – Image Processing, Computer Vision, Pattern Recognition, and Graphics

This Springer imprint is published by the registered company Springer Nature Switzerland AG
The registered company address is: Gewerbestrasse 11, 6330 Cham, Switzerland

Preface

This volume contains the papers presented at the 12th IAPR-TC15 Workshop on Graph-Based Representations in Pattern Recognition (GbR) held during June 19–21, 2019, in Tours.

In total 22 papers were accepted and presented orally. Each submission was reviewed by at least two and usually three Program Committee members. The program also included two very interesting invited talks: one by Christine Solnon, from the INSA of Lyon, who presented a talk entitled "Experimental Evaluation of Subgraph Isomorphism Solvers"; one by Marco Gori, from the University of Siena, who presented a talk entitled "Local Propagation in Graphical Neural Networks."

Accepted papers mainly cover the following topics: graph edit distance, graph matching, machine learning for graph problems, network and graph embedding, spectral graph problems, and parallel algorithms for graph problems. Numerous applications have been addressed with the help of graph-based representations, ranging from fMRI applications, image and video processing, to social networks analysis, document analysis, chemio-informatics and classification problems.

Authors of selected papers were invited to submit an extended version to a Special Issue on "Advances in Graph-based Representations for Pattern Recognition" to be published in *Pattern Recognition Letters* in 2020.

The GbR 2019 workshop was hosted by the Computer Science Laboratory of University of Tours in France (LIFAT). We acknowledge the generous support from the city of Tours, the French Region Centre Val de Loire, the University of Tours and the Engineering School of the University, the research federation ICVL, and the company APSIDE. We would like to thank all the Program Committee members for their help in the review process. We also wish to thank all the local organizers. Without their contributions, GbR 2019 would not have been successful. Finally, we express our appreciation to Springer for publishing this volume.

June 2019

Donatello Conte
Jean-Yves Ramel
Pasquale Foggia

Organization

Program Committee

Sebastien Bougleux	Normandie University, UNICAEN, ENSICAEN, CNRS, GREYC, France
Luc Brun	Normandie University, UNICAEN, ENSICAEN, CNRS, GREYC, France
Ananda S. Chowdhury	Jadavpur University, India
Donatello Conte	Computer Science Laboratory (LIFAT EA 6300), Tours, France
Guillaume Damiand	CNRS/LIRIS/Université de Lyon, France
Francisco Escolano	University of Alicante, Spain
Pasquale Foggia	University of Salerno, Italy
Benoit Gaüzère	Normandie Université, INSA de Rouen, LITIS, France
Giuliano Grossi	University of Milan, Italy
Edwin Hancock	University of York, UK
Pierre Héroux	Université de Rouen - LITIS EA 4108, France
Xiaoyi Jiang	University of Münster, Germany
Walter G. Kropatsch	Vienna University of Technology, Austria
Cheng-Lin Liu	Institute of Automation of Chinese Academy of Sciences, China
Josep Llados	Computer Vision Center, Universitat Autònoma de Barcelona, Spain
Bin Luo	Anhui University, China
Jean-Marc Ogier	University of La Rochelle, Laboratoire L3i, France
Marcello Pelillo	University of Venice, Italy
Jean-Yves Ramel	Computer Science Laboratory (LIFAT EA 6300), Tours, France
Romain Raveaux	Computer Science Laboratory (LIFAT EA 6300), Tours, France
Kaspar Riesen	University of Applied Sciences and Arts Northwestern Switzerland
Francesc Serratosa	Universitat Rovira i Virgili, Spain
Christine Solnon	LIRIS CNRS UMR 5205/INSA Lyon, France
Salvatore Tabbone	Université de Lorraine, France
Andrea Torsello	Università Ca Foscari, Italy
Ernest Valveny	Computer Vision Center - Universitat Autònoma de Barcelona, Spain
Mario Vento	Università degli Studi di Salerno, Italy
Richard Wilson	University of York, UK

Additional Reviewers

Frasca, Marco
Jin, Xiao-Bo
Lin, Jianyi
Wang, Xiao
Zhang, Yanming
Zhu, Yabin

Contents

Experimental Evaluation of Subgraph Isomorphism Solvers

Christine Solnon[✉]

INSA-Lyon, LIRIS UMR5205, 69621 Villeurbanne, France
christine.solnon@insa-lyon.fr

Abstract. Subgraph Isomorphism (SI) is an NP-complete problem
which is at the heart of many structural pattern recognition tasks as
it involves finding a copy of a pattern graph into a target graph. In
the pattern recognition community, the most well-known SI solvers are
VF2, VF3, and RI. SI is also widely studied in the constraint programming community, and many constraint-based SI solvers have been proposed since Ullman, such as LAD and Glasgow, for example. All these
SI solvers can solve very quickly some large SI instances, that involve
graphs with thousands of nodes. However, McCreesh *et al.* have recently
shown how to randomly generate SI instances the hardness of which can
be controlled and predicted, and they have built small instances which
are computationally challenging for all solvers. They have also shown
that some small instances, which are predicted to be easy and are easily solved by constraint-based solvers, appear to be challenging for VF2
and VF3. In this paper, we widen this study by considering a large test
suite coming from eight benchmarks. We show that, as expected for an
NP-complete problem, the solving time of an instance does not depend
on its size, and that some small instances coming from real applications
are not solved by any of the considered solvers. We also show that, if
RI and VF3 can solve very quickly a large number of easy instances, for
which Glasgow or LAD need more time, they fail at solving some other
instances that are quickly solved by Glasgow or LAD, and they are clearly
outperformed by Glasgow on hard instances. Finally, we show that we
can easily combine solvers to take benefit of their complementarity.

1 Introduction

Subgraph Isomorphism (SI) is an NP-complete problem which involves finding
a copy of a pattern graph into a target graph, *i.e.*, finding a mapping that
associates a different target node to each pattern node in such a way that edges
are preserved. There are two main variants of SI: in the non-induced case, only
pattern edges must be preserved (*i.e.*, pattern nodes connected by an edge must

---

This work has been done in collaboration with Ciaran McCreesh, Patrick Prosser, and
James Trimble. In particular, all experiments have been run by Ciaran McCreesh and
used the Cirrus UK National Tier-2 HPC Service at EPCC (http://www.cirrus.ac.uk)
funded by the University of Edinburgh and EPSRC (EP/P020267/1).

D. Conte et al. (Eds.): GbRPR 2019, LNCS 11510, pp. 1–13, 2019.
https://doi.org/10.1007/978-3-030-20081-7_1

be mapped to target nodes connected by an edge); in the induced case, target edges must also be preserved (*i.e.*, target nodes connected by an edge cannot be mapped to pattern nodes not connected by an edge).

SI is at the heart of many structural pattern recognition tasks in different application fields such as image or biology, for example [7]. In the pattern recognition community, the most well-known algorithms used to solve SI are VF2 [8], VF3 [5], and RI [4]. These solvers will be referred to as *PR solvers*. PR solvers perform a depth-first search in a space of states: each state corresponds to a partial mapping where some pattern nodes have been mapped, and each state is recursively extended by adding to its partial mapping a new couple of mapped pattern/target nodes.

SI is also widely studied in the constraint programming community as it may be modelled as a constraint satisfaction problem in a straightforward way. Many constraint-based solvers have been proposed for solving SI since Ullman [20] such as, for example, nRF+ [15], ILF [21], LAD [18], SND [2], and Glasgow [1,16]. These solvers will be referred to as *CP solvers*. Like VF2, VF3, and RI, CP solvers recursively extend partial mappings. However, a fundamental difference is that CP solvers maintain, for each non-mapped pattern node, the list of candidate target nodes that may be mapped to it, and they propagate constraints to reduce these lists. This constraint propagation mechanism is expensive, both in memory and time, but it reduces the number of states to explore.

Recent PR and CP solvers can solve very quickly rather large SI instances, that involve graphs with thousands of nodes. Indeed, being NP-complete does not mean that all instances are hard to solve, and some instances of NP-complete problems can be very easy to solve. In particular, in [6], Cheeseman *et al.* show that NP-complete problems can be summarised by at least one "order parameter", and that hard instances occur at a critical value of such a parameter. In [17], McCreesh *et al.* use this approach to generate "really hard" random SI instances according to three random graph models. For example, for *Erdős-Rényi* random graphs (where edges are generated according to an independent probability [11]), instances of non-induced SI may be generated by fixing pattern and target numbers of nodes, and varying pattern and target edge probabilities from 0 to 1. In this case, a phase transition occurs between entirely satisfiable instances (when patterns are sparse and targets are dense) and entirely unsatisfiable instances (when patterns are dense and targets are sparse), and the location of this phase transition can be predicted by computing the expected number of solutions. Instances located within this phase transition are computationally challenging for all solvers even when graphs are small (*e.g.*, thirty pattern nodes and 150 target nodes). However, the experimental study reported in [17] also shows that some small instances which are predicted as easy, and which are easily solved by CP solvers, appear to be challenging for PR solvers.

In this paper, we widen this experimental study and we experimentally evaluate and compare RI, VF2, VF3, Glasgow, and LAD on a large test suite of 14,621 instances coming from eight benchmarks. In Sect. 2, we describe our test suite. In Sect. 3, we show that, as expected for an NP-complete problem, the solving time of an instance does not depend on its size, and that some small instances (including instances coming from real applications) are not solved by any of the considered solvers. In Sect. 4, we identify easy and hard instances and we show that, if PR solvers are able to solve very quickly easy instances (for which CP solvers often need more time), they fail at solving some other instances that are rather quickly solved by CP solvers, and they are clearly outperformed by Glasgow on hard instances. Finally, in Sect. 5, we show that we can easily combine PR and CP solvers to take benefit of their complementarity.

Table 1. For each class, we give the number of instances (#inst) and then describe pattern and target graph features: minimum and maximum number of nodes, number of edges, and density.

Class	#inst	Pattern graphs						Target graphs					
		#nodes		#edges		Density		#nodes		#edges		Density	
		min	max	min	max	min	max	min	max	min	max	min	max
images	6,302	4	170	4	241	.02	.67	1,072	5,972	1,539	8,888	.00	.00
meshes	3,018	40	199	114	539	.02	.15	201	5,873	252	15,292	.00	.02
LV	3,831	10	128	10	4,950	.02	1.00	10	6,671	10	209,000	.00	1.00
randERP	200	30	30	128	387	.29	.89	150	150	4,132	8,740	.37	.78
randER	270	40	360	41	12,410	.02	.21	200	600	436	34,210	.02	.19
randBVG	540	40	480	43	2,137	.01	.20	200	800	299	3,600	.00	.05
randM	360	51	777	76	2,075	.01	.08	256	1,296	672	4,377	.00	.03
randSF	100	180	900	478	5,978	.01	.17	200	1000	592	7,148	.01	.16

2   Experimental Set-Up

Test Suite. We consider 14, 621 instances coming from eight benchmarks described in Table 1, and available at liris.cnrs.fr/christine.solnon/SIP.html. *images* and *meshes* are coming from real applications where both pattern and target graphs correspond to graphs extracted from segmented images and 3D meshes [9,19].

LV is a benchmark described in [15]. It uses 113 graphs with various properties coming from the Stanford GraphBase described by Knuth in [13]. The benchmark is built by splitting the set of graphs in two parts: the first part contains the 50 smallest graphs; the second part contains the 63 remaining graphs. We consider all pairs of graphs such that the pattern graph belongs to the first part, the target graph belongs to the first or the second part, and the target graph has at least as many nodes as the pattern graph.

rand* (with $* \in \{ERP, ER, BVG, M, SF\}$) are randomly generated instances. *randERP* are instances close to the phase transition (expected to be hard as explained in [17]), and all graphs are *Erdős-Rényi* graphs. *randER*, *randBVG*, and *randM* are coming from the database described in [10], and graphs are *Erdős-Rényi* graphs, (modified) bounded valence graphs and 4D meshes, respectively.

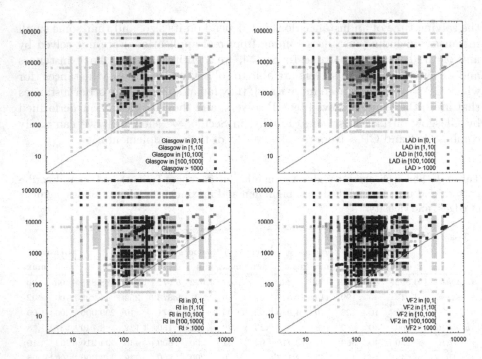

Fig. 1. Number of pattern edges (x-axis), target edges (y-axis), and solving time (colour) for non-induced SI: top left = Glasgow; top right = LAD; bottom left = RI; bottom right = VF2. (Color figure online)

randSF is described in [22] and it contains scale-free graphs. All instances in *randER*, *randBVG*, *randM*, and *randSF* (except 20 instances in *randSF*) are feasible by construction because the pattern has been extracted from the target.

All graphs have at least as many edges as nodes. Hence, the size of a graph is dominated by its number of edges.

Performance Measures. The experiments were performed on the EPCC Cirrus HPC facility, on systems with dual Intel Xeon E5-2695 v4 CPUs and 256 Gb RAM, running Centos 7.3.1611, and GCC 7.2.0 as the compiler. Each run has been limited to 1,000 s of CPU time. Some instances are not solved within this time limit (note that even when increasing the time limit to 100,000 s some instances are still unsolved). We consider two different performance measures: when all solvers have been able to solve all instances of a benchmark, we report the average solving time; when some instances have not been solved within the time limit, we report the number of solved instances within the time limit, and we plot the evolution of the cumulative number of solved instances with respect to time (*i.e.*, the function $f(t) = \#\{i \in I : t_i^s \le t\}$ where I is the set of instances, s a solver, and t_i^s the time spent by s to solve an instance $i \in I$).

We do not consider memory consumption as a performance measure as solvers never run out of memory, even for the largest instances (all solvers have polynomial memory complexities). However, CP solvers need more memory than PR solvers as they maintain candidate lists of target nodes for each non-mapped pattern vertex.

Different variants of Glasgow are described in [1]. We consider the *biased* variant, which is the default setting[1].

3 Does the Solving Time Depend on Graph Sizes?

To study the relation between the solving time and the size of an instance, we plot in Figs. 1 and 2 the time spent by each solver on each instance. Each instance corresponds to a point (x, y) where x is the number of pattern edges, y the number of target edges, and the colour depends on the solving time: yellow if it is smaller than one second, and black if the instance has not been solved within 1000 s (if several instances have the same size, the colour corresponds to the average solving time for all these instances).

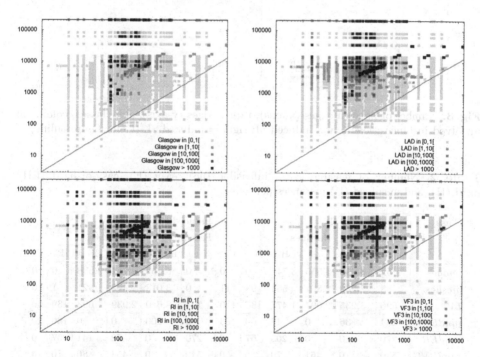

Fig. 2. Number of pattern edges (x-axis), target edges (y-axis), and solving time (colour) for induced SI: top left = Glasgow; top right = LAD; bottom left = RI; bottom right = VF3. (Color figure online)

---

[1] Glasgow is available at https://github.com/ciaranm/glasgow-subgraph-solver.

As expected for an NP-complete problem, these figures show us that *hardness does not depend on size*. Let us first consider the non-induced case, displayed in Fig. 1. Unsolved instances (black points) are not specially concentrated in the top right area of the plots (corresponding to the largest instances). The number of unsolved instances is quite different from a solver to another, but some black points are common to all solvers. Among the set of instances which are solved by none of the solvers, the smallest pattern (resp. target) graph has 62 edges and 30 nodes (resp. 400 edges and 86 nodes). Many much larger instances are solved in less than one second. The gray line separates instances that have more target edges than pattern edges (top left) from those that have less target edges than pattern edges (bottom right). All instances in the bottom right part are trivially infeasible. However, both VF2 and RI are not able to solve some of them.

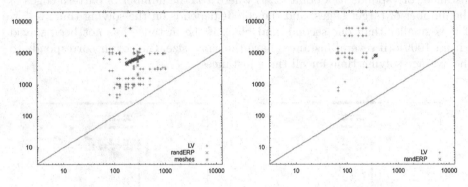

Fig. 3. Number of pattern edges (x-axis), target edges (y-axis), and classes (colour) of unsolved instances: left = non-induced SI; right = induced SI. (Color figure online)

Table 2. Number of feasible (yes), infeasible (no), easy (E), easy-or-hard (EH), hard (H), and unsolved (U) instances per class.

Class	Non-induced SI						Induced SI					
	Feasibility		Hardness				Feasibility		Hardness			
	yes	no	E	EH	H	U	yes	no	E	EH	H	U
images	52	6,250	2,555	3,747	0	0	50	6,252	2,764	3,538	0	0
meshes	88	2,930	2,361	553	93	11	0	3,018	2,492	521	1	0
LV	596	3,235	2,097	1,477	137	120	191	3,640	2,939	693	139	60
randERP	164	36	0	48	69	83	0	200	0	0	180	20
randER	270	0	0	203	67	0	270	0	72	141	57	0
randBVG	540	0	461	79	0	0	540	0	454	86	0	0
randM	360	0	309	51	0	0	360	0	313	45	2	0
randSF	80	20	4	96	0	0	80	20	75	25	0	0
All	2,150	12,471	7,787	6,254	366	214	1,491	13,130	9,109	5,053	379	80

When looking at the induced case in Fig. 2, we also note that the unsolved instances are not necessarily those with the largest graphs and the number of unsolved instances is quite different from a solver to another. Among the set of instances which are solved by none of the solvers, the smallest pattern (resp. target) graph has 62 edges and 30 nodes (resp. 638 edges and 120 nodes). VF3 has much better results on induced SI than VF2 on non-induced SI, and it is always able to quickly solve instances that are trivially infeasible because they have less target edges than pattern edges.

4   Where Are the Hard Instances?

To have a better insight into where the hard instances are, we have partitioned each class of our benchmark into 4 separate groups, depending on instance hardness. As all instances but those of *randERP* have not been randomly generated with a model that allows us to predict hardness with respect to the phase transition location, we consider an empirical definition of instance hardness:

- an instance is *easy* if the four solvers are able to solve it within one second;
- an instance is *hard* if no solver can solve it within one second, but at least one solver can solve it within the time limit of 1000 s;
- an instance is *easy-or-hard* if at least one solver solves it within one second whereas at least one solver cannot solve it within one second;
- an instance is *unsolved* if no solver can solve it within the time limit of 1000 s.

In Fig. 3, we display the number of edges in pattern and target graphs of unsolved instances, and in Table 2, we give the number of instances in each group of each class. As expected, many *randERP* instances are unsolved or hard, and none of them is easy: these instances are close to the phase transition and they are expected to be challenging despite their small size. However, not all unsolved instances are coming from *randERP*. This shows us that really hard instances may occur even if they have not been generated on purpose. For the non-induced case, *LV* and *meshes* respectively contain 120 and 11 unsolved instances, whereas for the induced case, *LV* contains 60 unsolved instances. In both cases, these instances are not the largest ones, and some of them are really small as illustrated in Fig. 3.

Many instances are easy (7,787 instances for the non-induced case, and 9,109 for the induced case), and these easy instances are coming from all classes but *randERP* and *randER* for the non-induced case, and all classes but *randERP* for the induced case.

In Table 2, we also give the number of feasible instance per class. Note that any instance feasible for the induced case is also feasible for the non-induced case. Three classes (*i.e.*, *randER*, *randBVG*, and *randM*) only contain feasible instances as they have been randomly generated in such a way that there always exists at least one solution. There is no obvious relation between feasibility and hardness: hard and unsolved groups contain both feasible and infeasible instances.

Table 3. Results of Glasgow (G), LAD (L), VF2/VF3 (V), and RI (R) on non-induced (top) and induced (bottom) SI instances. #u is the number of unsolved instances within 1000 s (for easy instances, #u = 0). When all instances are solved, we report the average solving time in seconds.

	easy instances				easy-or-hard instances								hard instances			
	G	L	V	R	G		L		V		R		G	L	V	R
Class	time	time	time	time	#u	time	#u	time	#u	time	#u	time	#u	#u	#u	#u
								Non-induced SI								
images	.106	.374	.201	.002	0	(.18)	0	(4.07)	13	-	0	(0.01)	-	-	-	-
meshes	.153	.026	.044	.016	1	-	3	-	276	-	180	-	20	23	91	91
LV	.036	.017	.069	.008	12	-	9	-	886	-	206	-	18	32	130	76
randERP	-	-	-	-	0	(.21)	25	-	48	-	30	-	0	65	69	64
randER	-	-	-	-	0	(.88)	17	-	201	-	2	-	0	57	67	14
randBVG	.017	.119	.004	.003	0	(.07)	0	(2.37)	0	(.01)	2	-	-	-	-	-
randM	.051	.095	.013	.003	0	(.18)	0	(3.94)	18	-	2	-	-	-	-	-
randSF	.007	.004	.497	.001	0	(.11)	0	(.10)	80	-	15	-	-	-	-	-
All	.094	.146	.099	.008	13	-	54	-	1,522	-	437	-	38	177	357	245
								Induced SI								
images	.136	.388	.002	.002	0	(.24)	0	(3.58)	0	(.00)	0	(.01)	-	-	-	-
meshes	.173	.024	.001	.009	0	(1.04)	0	(.16)	103	-	237	-	0	0	1	1
LV	.047	.026	.011	.024	4	-	8	-	52	-	96	-	10	53	51	65
randERP	-	-	-	-	-	-	-	-	-	-	-	-	0	175	175	175
randER	.021	.260	.061	.030	0	(.43)	12	-	2	-	1	-	0	47	5	7
randBVG	.018	.125	.002	.004	0	(.07)	0	(2.22)	3	-	4	-	-	-	-	-
randM	.052	.117	.003	.003	0	(.17)	0	(4.31)	6	-	0	(6.11)	0	0	1	0
randSF	.109	.065	.027	.023	0	(.14)	0	(.21)	2	-	10	-	-	-	-	-
All	.108	.146	.005	.012	4	-	20	-	168	-	348	-	10	275	233	248

5   Experimental Comparison of the Solvers

In Table 3, we display the results of the four solvers on the different classes, grouped with respect to hardness. For easy instances (which are solved by all solvers), RI is an order faster than the other solvers for the non-induced case, and VF3 is twice as fast as RI which is an order faster than Glasgow and LAD for the induced case. Hence, on easy instances, the fastest solvers clearly are RI for the non-induced case and VF3 for the induced case, and CP solvers are an order slower.

However, on easy-or-hard and hard instances, PR solvers solve less instances than LAD, and LAD solves less instances than Glasgow. More precisely, for the non-induced case, Glasgow (resp. LAD, VF2, and RI) fails at solving 51 (resp. 221, 1879, and 682) instances. For the induced case, Glasgow (resp. LAD, VF3, and RI) fails at solving 14 (resp. 295, 416, and 596) instances. Hence, on easy-or-hard and on hard instances, the best solver clearly is Glasgow for both the non-induced and the induced case. Actually most easy-or-hard instances are trivially solved by Glasgow in less than one second whereas PR solvers fail at solving many of these instances.

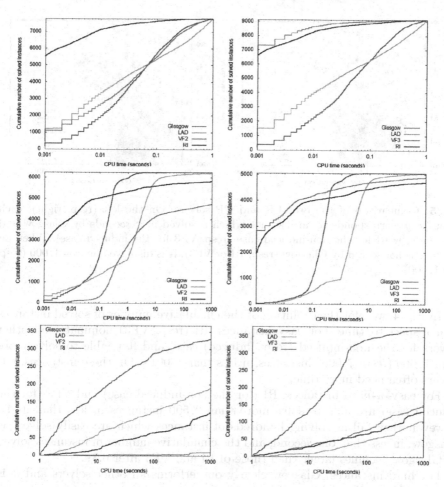

Fig. 4. Cumulative number of solved instances: top = easy instances; middle = easy-or-hard instances; bottom = hard instances; left = non-induced SI; right = induced SI.

For the non-induced case, if LAD is outperformed by Glasgow, it is able to solve much more instances than PR solvers. For the induced case, LAD is also outperformed by Glasgow and, if it is able to solve more instances than PR solvers on many classes, it is clearly outperformed by them on *randER* instances. Actually, LAD is the only solver which solves less instances for the induced case than for the non-induced case. This comes from the fact that LAD has been designed for the non-induced case. It has been extended to handle the induced case in a very naive way (by checking that target edges are preserved *a posteriori*), without exploiting properties specific to the induced case.

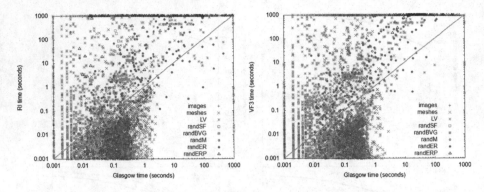

Fig. 5. Comparison of the best PR and CP solvers. On the left (resp. right), each point (x, y) corresponds to an instance which is solved in x seconds by Glasgow and y seconds by RI for the non-induced case (resp. VF3 for the induced case). When an instance is not solved by Glasgow (resp. RI or VF3), it is displayed on $x = 1,000$ (resp. $y = 1,000$).

In Fig. 4, we plot the evolution of the cumulative number of solved instances with respect to time. For easy instances, RI (resp. VF3) dominates all other solvers for the non-induced (resp. induced) case, and it is able to solve more than $5,000$ (resp. $7,000$) instances in less than 0.001 s. On these instances, CP solvers often need more time.

For easy-or-hard instances, RI (for the non-induced case) and VF3 (for the induced case) are able to solve more than $2,500$ instances in less than .001 s. However, they fail at solving hundreds of instances which are easily solved by Glasgow, in less than one second, and the cumulative number of instances solved by Glasgow becomes larger than those of RI and VF3 after 0.3 s.

For hard instances, Glasgow clearly outperforms all other solvers and it is able to solve much more instances.

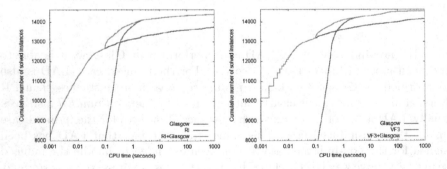

Fig. 6. Cumulative number of solved instances on the whole benchmark of RI (resp. VF3), Glasgow, and RI+Glasgow (resp. VF3+Glasgow) for non-induced SI (left) (resp. induced SI (right)).

In Fig. 5, we compare the best CP solver (*i.e.*, Glasgow) with the best PR solver (*i.e.*, RI for the non-induced case, and VF3 for the induced case) on a per instance basis. Every point below the gray line corresponds to an instance which is solved quicker by the PR solver than by Glasgow, and the wide majority of these points are on the left of the vertical line $x = 1$, corresponding to instances which are solved in less than one second by Glasgow. Every point above the gray line corresponds to an instance which is solved quicker by Glasgow than by the PR solver, and many of these points are on the horizontal line $y = 1,000$, corresponding to instances which are not solved by the PR solver within the time limit of $1,000$ s.

6   Combining Solvers to Take the Best of Them

Glasgow is complementary to the best PR solver (*i.e.*, RI for the non-induced case and VF3 for the induced case) as it needs more time on very easy instances, but it is able to solve more instances. We can take benefit of this complementarity as follows: we run the best PR solver with a time limit of t_1 seconds; if the instance has not been solved within this limit, we run Glasgow. The time limit t_1 should be long enough to allow the PR solver to solve easy instances, but not too long in order not to penalise the total solving time when the PR solver is not able to solve the instance. In Fig. 6, we display cumulative numbers of solved instances of the best PR solver, Glasgow, and the combined approach (denoted RI+Glasgow for the non-induced case, and VF3+Glasgow for the induced case) when the time limit t_1 is set to 0.1 s. It shows us that this simple combination allows to take the best of both solvers: before 0.1 s, the cumulative number of solved instances of RI+Glasgow (or VF3+Glasgow) is equal to the one of RI (or VF3), which is much greater than the one of Glasgow (not displayed because the y-axis starts at 8,000 and Glasgow solves less than 8,000 instances in 0.1 s); after 0.1 s, the cumulative number of solved instances of RI+Glasgow (or VF3+Glasgow) grows faster than the one of RI (or VF3) because Glasgow is able to solve instances which are not solved by RI (or VF3); finally, after a few seconds, the cumulative number of solved instances of RI+Glasgow (or VF3+Glasgow) is very close to the one of Glasgow as the delay of 0.1 s due to the run of RI (or VF3) is negligible.

Of course, this very simple approach could be enhanced by considering more solvers (including more variants of each solver, using different ordering heuristics, for example). In this case, we may gather all solvers in a portfolio, and use an algorithm selection approach to dynamically select from the portfolio the solver which is expected to perform best for each new SI instance to solve, as proposed by Kotthoff *et al.* in [14].

7   Conclusion

This study has shown that there are many very easy SI instances which are solved in a few milliseconds by modern solvers, and that some of these instances may involve very large graphs with thousands of nodes. However, there are still small instances which cannot be solved within a reasonable amount of time by any of these solvers. It is important to evaluate solvers on these hard instances too as they do appear in real applications, though they are less frequent than easy instances.

A promising research direction for solving hard instances is to exploit multiple cores, and parallel SI solvers have been introduced in [1,3,16], for example. A special attention should be paid on performance measures used to evaluate these approaches. Indeed, measuring an average speed-up between a sequential and a parallel solver is not very meaningful when considering NP-complete problems because speed-ups are very different from an instance to another, and do not depend on instance sizes: for easy instances, speed-ups are usually very low, whereas for hard instances it is not rare to have super-linear speed-ups. Also, really hard instances are not solved within a reasonable amount of time, and speed-ups cannot be computed in this case. Let us illustrate this point on the parallel version of Glasgow (using 32 cores) described in [1]. On easy instances (solved in less than 1 s by sequential Glasgow), the speed-up varies between 0.1 and 32, and the average speed-up is close to 1. On hard instances (that are not solved by sequential Glasgow within 1 s, but are solved within 1000 s), the speed-up varies between 1 and 583, and the average speed-up is 14. However, parallel Glasgow is able to solve instances which are not solved by sequential Glasgow within 1000 s and, if we include these instances, the average speed-up becomes greater than 19 (this is a lower bound of the speed-up as we only have a lower bound of the time of sequential Glasgow for unsolved instances). This shows us that the average speed-up does not give a clear picture of solver performance. Better insights are given by scatter plots that compare times on a per instance basis (as done in Fig. 5), or by the aggregate speed-up measure introduced in [12], which measures timeout ratio for solving a same number of instances. For instance, Sequential Glasgow solves 14, 356 instances within 1000 s, and the hardest of these instances is solved in 939 s. Parallel Glasgow solves 14, 356 instances within a timeout of 19 s, and this gives an aggregate speed-up of $939/19 = 49$.

References

1. Archibald, B., Dunlop, F., Hoffmann, R., McCreesh, C., Prosser, P., Trimble, J.: Sequential and parallel solution-biased search for subgraph algorithms. In: 16th International Conference on Integration of Constraint Programming, Artificial Intelligence, and Operations Research (2019)
2. Audemard, G., Lecoutre, C., Samy-Modeliar, M., Goncalves, G., Porumbel, D.: Scoring-based neighborhood dominance for the subgraph isomorphism problem. In: O'Sullivan, B. (ed.) CP 2014. LNCS, vol. 8656, pp. 125–141. Springer, Cham (2014). https://doi.org/10.1007/978-3-319-10428-7_12
3. Bombieri, N., Bonnici, V., Giugno, R.: Parallel searching on biological networks. In: 27th Euromicro International Conference on Parallel, Distributed and Network-Based Processing, PDP, pp. 307–314. IEEE (2019)
4. Bonnici, V., Giugno, R.: On the variable ordering in subgraph isomorphism algorithms. IEEE/ACM Trans. Comput. Biol. Bioinf. **14**(1), 193–203 (2017)

5. Carletti, V., Foggia, P., Saggese, A., Vento, M.: Challenging the time complexity of exact subgraph isomorphism for huge and dense graphs with VF3. IEEE Trans. Pattern Anal. Mach. Intell. **40**(4), 804–818 (2018)
6. Cheeseman, P., Kanefsky, B., Taylor, W.M.: Where the really hard problems are. In: 12th International Joint Conference on Artificial Intelligence (IJCAI), pp. 331–340 (1991)
7. Conte, D., Foggia, P., Sansone, C., Vento, M.: Thirty years of graph matching in pattern recognition. IJPRAI **18**(3), 265–298 (2004)
8. Cordella, L.P., Foggia, P., Sansone, C., Vento, M.: A (sub)graph isomorphism algorithm for matching large graphs. IEEE Trans. Pattern Anal. Mach. Intell. **26**(10), 1367–1372 (2004)
9. Damiand, G., Solnon, C., de la Higuera, C., Janodet, J.C., Samuel, E.: Polynomial algorithms for subisomorphism of nD open combinatorial maps. Comput. Vis. Image Underst. (CVIU) **115**(7), 996–1010 (2011)
10. De Santo, M., Foggia, P., Sansone, C., Vento, M.: A large database of graphs and its use for benchmarking graph isomorphism algorithms. Pattern Recogn. Lett. **24**(8), 1067–1079 (2003)
11. Erdős, P., Rényi, A.: On random graphs I. Publicationes Mathematicae **6**, 290–297 (1959)
12. Hoffmann, R., et al.: Observations from parallelising three maximum common (connected) subgraph algorithms. In: van Hoeve, W.-J. (ed.) CPAIOR 2018. LNCS, vol. 10848, pp. 298–315. Springer, Cham (2018). https://doi.org/10.1007/978-3-319-93031-2_22
13. Knuth, D.E.: The Stanford GraphBase - a platform for combinatorial computing. ACM (1993)
14. Kotthoff, L., McCreesh, C., Solnon, C.: Portfolios of subgraph isomorphism algorithms. In: Festa, P., Sellmann, M., Vanschoren, J. (eds.) LION 2016. LNCS, vol. 10079, pp. 107–122. Springer, Cham (2016). https://doi.org/10.1007/978-3-319-50349-3_8
15. Larrosa, J., Valiente, G.: Constraint satisfaction algorithms for graph pattern matching. Math. Struct. Comput. Sci. **12**(4), 403–422 (2002)
16. McCreesh, C., Prosser, P.: A parallel, backjumping subgraph isomorphism algorithm using supplemental graphs. In: Pesant, G. (ed.) CP 2015. LNCS, vol. 9255, pp. 295–312. Springer, Cham (2015). https://doi.org/10.1007/978-3-319-23219-5_21
17. Mccreesh, C., Prosser, P., Solnon, C., Trimble, J.: When subgraph isomorphism is really hard, and why this matters for graph databases. J. Artif. Intell. Res. **61**, 723–759 (2018)
18. Solnon, C.: AllDifferent-based filtering for subgraph isomorphism. Artif. Intell. **174**(12–13), 850–864 (2010)
19. Solnon, C., Damiand, G., de la Higuera, C., Janodet, J.: On the complexity of submap isomorphism and maximum common submap problems. Pattern Recogn. **48**(2), 302–316 (2015)
20. Ullmann, J.R.: An algorithm for subgraph isomorphism. J. ACM **23**(1), 31–42 (1976)
21. Zampelli, S., Deville, Y., Solnon, C.: Solving subgraph isomorphism problems with constraint programming. Constraints **15**(3), 327–353 (2010)
22. Zampelli, S., Deville, Y., Solnon, C., Sorlin, S., Dupont, P.: Filtering for subgraph isomorphism. In: Bessière, C. (ed.) CP 2007. LNCS, vol. 4741, pp. 728–742. Springer, Heidelberg (2007). https://doi.org/10.1007/978-3-540-74970-7_51

GEDLIB: A C++ Library for Graph Edit Distance Computation

David B. Blumenthal[1]($\boxtimes$) ⓘ, Sébastien Bougleux[2] ⓘ, Johann Gamper[1] ⓘ,
and Luc Brun[2] ⓘ

[1] Faculty of Computer Science, Free University of Bozen-Bolzano, Bolzano, Italy
{david.blumenthal,gamper}@inf.unibz.it
[2] Normandie Univ, UNICAEN, ENSICAEN, CNRS, GREYC, Caen, France
bougleux@unicaen.fr, luc.brun@ensicaen.fr

Abstract. The graph edit distance (GED) is a flexible graph dissimilarity measure widely used within the structural pattern recognition field. In this paper, we present GEDLIB, a C++ library for exactly or approximately computing GED. Many existing algorithms for GED are already implemented in GEDLIB. Moreover, GEDLIB is designed to be easily extensible: for implementing new edit cost functions and GED algorithms, it suffices to implement abstract classes contained in the library. For implementing these extensions, the user has access to a wide range of utilities, such as deep neural networks, support vector machines, mixed integer linear programming solvers, a blackbox optimizer, and solvers for the linear sum assignment problem with and without error-correction.

Keywords: Graph edit distance · Open source library · C++

1 Introduction

Because of their expressiveness and versatility, labeled graphs are widely used to model various kinds of objects such as molecules, street networks, and images. Many pattern recognition problems defined over these domains presuppose the availability of a (dis-)similarity measure for labeled graphs. Despite the fact that its exact computation is $\mathcal{NP}$-hard [31], one of the most widely used measures is the *graph edit distance* (GED). Given two labeled graphs G and H, it is defined as $\mathrm{GED}(G, H) := \min_{P \in \Psi(G,H)} c(P)$, where Ψ is the set of all *edit paths* between G and H and $c(P)$ denotes the cost of an edit path P. An edit path is a sequence of *edit operations* that transforms G into H. There are six edit operations: substituting a node or an edge in G by a node or an edge in H, deleting an edge or an isolated node from G, and inserting an edge or an isolated node into H. Each edit operation comes with an associated non-negative *edit cost* defined in terms of the node or edge labels involved in the operation; and the cost of an edit path is defined as the sum over the costs of its edit operations.

Over the past years, some exact and a lot of approximate algorithms for computing GED have been suggested. As the hardness of GED does not allow

ⓒ Springer Nature Switzerland AG 2019
D. Conte et al. (Eds.): GbRPR 2019, LNCS 11510, pp. 14–24, 2019.
https://doi.org/10.1007/978-3-030-20081-7_2

for a theoretical evaluation of approximate algorithms (the existence of any α-approximation algorithm for GED would imply that the graph isomorphism problem, a prime candidate for an $\mathcal{NP}$-intermediate problem, is in $\mathcal{P}$), these algorithms are typically evaluated empirically. In order for such a comparison to be fair, it is highly desirable that the compared algorithms be implemented within the same environment. However, to the best of our knowledge, no software is available that can be used for this purpose.

In this paper, we present the C++ template library GEDLIB which is intended to fill this gap. GEDLIB is available on GitHub:

https://github.com/dbblumenthal/gedlib

In its current version, GEDLIB contains implementations of 24 different GED algorithms and 9 different edit cost functions. Further algorithms and edit costs can be implemented easily by implementing abstract classes contained in GEDLIB. For this, the user has access to standard libraries for blackbox optimization, mixed integer linear programming, the linear sum assignment problem with and without error-correction, deep neural networks, and support vector machines. GEDLIB provides a parser to load graphs given in the GXL file format. Alternatively, graphs with user-specified node ID, node, and edge label types can be constructed from within GEDLIB. Internally, GEDLIB uses the Boost Graph Library [22] for representing the graphs and Eigen [19] for matrix operations.

The remainder of this paper is organized as follows: In Sect. 2, the overall architecture of GEDLIB is sketched. In Sect. 3, the user interface is presented. In Sects. 4 and 5, the abstract classes for implementing GED algorithms and edit cost functions are described. Section 6 concludes the paper. Details, examples, and installation instructions can be found in the documentation.

2   Overall Architecture

Figure 1 shows the overall architecture of GEDLIB in a UML diagram. The entire library is contained in the namespace ged. The template parameters UserNodeID, UserNodeLabel, and UserEdgeLabel correspond to the types of the node IDs, the node labels, and the edge labels of the graphs provided by the user.

- The class template ged::GEDEnv provides the user interface. Via its public member functions, graphs can be constructed or loaded from GXL files, edit costs can be set, the algorithms implemented in GEDLIB can be run, and the results of the runs can be obtained. For users who do not want to provide extensions for GEDLIB, it suffices to get familiar with this class template.
- The abstract class template *ged::GEDMethod* provides a generic interface for implementing algorithms that exactly or approximately compute GED.
- The abstract class templates *ged::LSBasedMethod*, *ged::MIPBasedMethod*, and *ged::LSAPEBasedMethod* are derived from the generic interface provided by *ged::GEDMethod*. They yield more specialized interfaces for implementing

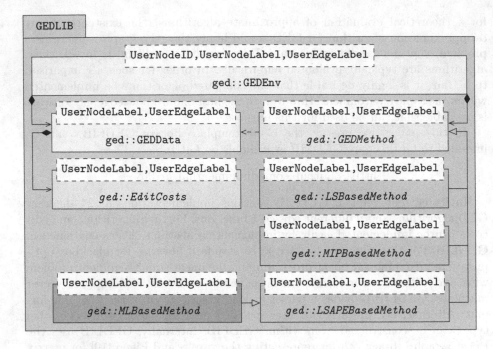

Fig. 1. The overall architecture of GEDLIB shown in a UML class diagram.

methods using local search, mixed integer linear programming, and transformations to the linear sum assignment problem with error-correction.

- The abstract class template *ged::MLBasedMethod* is derived from the interface *ged::LSAPEBasedMethod*. It can be used to implement algorithms that use deep neural networks or support vector machines for transforming GED to the linear sum assignment problem with error-correction.

- The class template ged::GEDData contains the normalized input data on which all GED algorithms contained in GEDLIB operate. Via the public member functions of ged::GEDData, derived classes of *ged::GEDMethod* have access to the graphs that have been added to the environment and to the edit cost functions selected by the user.

- The abstract class template *ged::EditCosts* provides a generic interface for implementing edit cost functions.

3   User Interface

In Fig. 2, the class template ged::GEDEnv, which constitutes the user interface of GEDLIB, is displayed in detail. By calling add_graph(), add_node(), and add_edge(), the user can add labeled graphs to the environment. Alternatively, load_gxl_graphs() can be used to load graphs given in the GXL file format. For this, the template parameter UserNodeID must be set to `ged::GXLNodeID` a.k.a. `std::string`, and the template parameters `UserNodeLabel` and `UserEdgeLabel` must be set to `ged::GXLLabel` a.k.a. `std::map<std::string,std::string>`.

Calls to `set_edit_costs()` add edit cost functions to the environment. The user can either select one of the predefined edit cost functions or use her own implementation of *ged::EditCosts*. Calls to `init()` initialize the environment eagerly or lazily. If eager initialization is chosen, all edit costs between graphs contained in the environment are precomputed. Otherwise, the edit cost functions are evaluated on the fly. The member function `set_method()` selects one of the GED algorithms available in GEDLIB. Some algorithms accept options, which can be passed to `set_method()` as a string of the form `"[--<option> <arg>] [...]"`. Calls to `init_method()` initialize the selected method for runs between graphs contained in the environment, and calls to `run_method()` run the method between two specified graphs. The results of the runs (lower and upper bounds, runtimes, etc.) can be accessed via various getter member function.

`UserNodeID,UserNodeLabel,UserEdgeLabel`	
`ged::GEDEnv`	
`...`	// *misc. variables*
+ `add_graph()`	// *adds a graph to the environment*
+ `add_node()`	// *adds a node to a previously added graph*
+ `add_edge()`	// *adds an edge to a previously added graph*
+ `load_gxl_graphs()`	// *loads graphs given as GXL files*
+ `set_edit_costs()`	// *selects the edit costs*
+ `init()`	// *initializes the environment*
+ `set_method()`	// *selects the GED method*
+ `init_method()`	// *initializes the selected GED method*
+ `run_method()`	// *runs the selected GED method*
`...`	// *misc. member functions*

Fig. 2. The user interface `ged::GEDEnv`.

4 Abstract Classes for Implementing GED Algorithms

Generic Interface. Figure 3 details the abstract class template *ged::GEDMethod*, which provides the generic interface for implementing GED. The interface is defined by the virtual member functions starting with the prefix *ged_*. We here describe only the most important virtual member functions; the remaining ones are detailed in the documentation: *ged_run_()* runs the method between two input graphs, *ged_init_()* initializes the methods for the graphs that have been

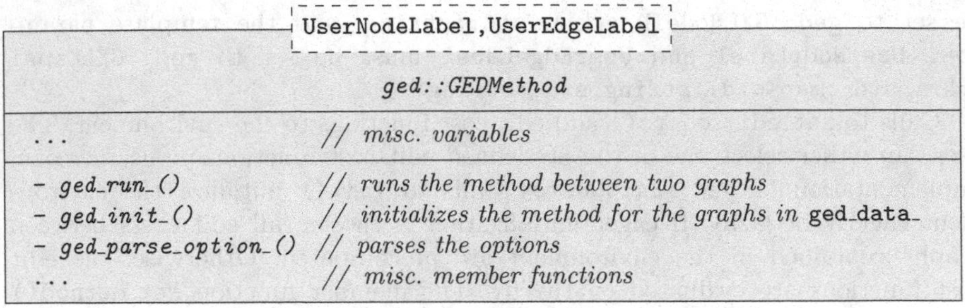

UserNodeLabel, UserEdgeLabel	
ged::GEDMethod	
...	// *misc. variables*
− *ged_run_()*	// *runs the method between two graphs*
− *ged_init_()*	// *initializes the method for the graphs in* ged_data_
− *ged_parse_option_()*	// *parses the options*
...	// *misc. member functions*

Fig. 3. The generic interface *ged::GEDMethod*.

added to the environment, and *ged_parse_option_()* parses the options of the method. The following existing algorithms already implemented in GEDLIB are directly derived classes of *ged::GEDMethod*: ged::BranchTight [2], ged::HED [17], ged::Partition [32], ged::Hybrid [32], ged::SimulatedAnnealing [30], ged::BranchCompact [32], ged::AnchorAwareGED [14].

Interface for Methods Based on the Linear Sum Assignment Problem with Error-Correction. A popular approach for approximating GED is to use transformations to the *linear sum assignment problem with error-correction* (LSAPE). An instance of LSAPE consists of a cost matrix $\mathbf{C} = (c_{i,k}) \in \mathbb{R}_{\geq 0}^{(n+1) \times (m+1)}$. The task is to compute a mapping π from rows to columns, such that each row except for $n + 1$ and each column expect for $m + 1$ is covered exactly once and $\mathbf{C}(\pi) \coloneqq \sum_{(i,k) \in \pi} c_{i,k}$ is minimized. LSAPE can be solved optimally in cubic time [10]; in GEDLIB, we use the LSAPE toolbox [8] for solving LSAPE.

If LSAPE is used for approximating GED(G, H), n and m are set to $|V^G|$ and $|V^H|$, the first $|V^G|$ rows of $\mathbf{C}$ are associated with the nodes of G, the first $|V^H|$ columns of $\mathbf{C}$ are associated with the nodes of H, and the last rows and columns are associated with dummy nodes used for codifying node insertions and deletions. With this setup, each LSAPE solution π corresponds to a *node map between G and H*, which, in turn, induces an edit path and hence an upper bound for GED(G, H) [6]. LSAPE based heuristics for GED try to achieve tight upper bounds by encoding structural information of the input graphs into $\mathbf{C}$. Moreover, some of them construct $\mathbf{C}$ such that $\min_\pi \mathbf{C}(\pi)$ lower bounds GED.

Figure 4 shows the abstract class template *ged::LSAPEBasedMethod*, which provides the interface for implementing heuristics of this kind. The interface is defined by the virtual member functions starting with the prefix *lsape_*. The most important one is *lsape_populate_instance_()*, which populates the LSAPE instance $\mathbf{C}$. The following algorithms implemented in GEDLIB are directly derived classes of *ged::LSAPEBasedMethod*: ged::Bipartite [26], ged::Branch [2], ged::BranchFast [2], ged::Node [21], ged::BranchUniform [32], ged::Ring [3], ged::Subgraph [12], ged::Walks [18]. Additionally, all derived classes of *ged::LSAPEBasedMethod* can be run with the node centralities suggested in [27].

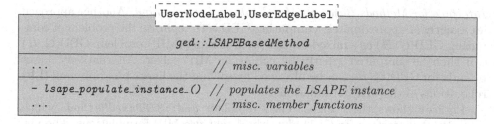

Fig. 4. The interface *ged::LSAPEBasedMethod* for methods based on LSAPE.

Interface for Methods Based on Machine Learning. Recently, it has been suggested to use deep neural networks or support vector machines for carrying out the transformation from GED to LSAPE. Given two graphs G and H, feature vectors are constructed for all node substitutions, deletions, and insertions, and the matrix $\mathbf{C}$ is defined as $c_{i,k} := 1 - p^*(i, k)$. Here, $p^*(i, k)$ is the confidence of a machine learning framework (either a deep neural network or a support vector machine) that the feature vector associated to the node edit operation corresponding to row i and column k is contained in an optimal node map.

Figure 5 details the abstract class template *ged::MLBasedMethod*, which provides the interface for algorithm adopting this paradigm. For implementing the interface, it suffices to override the virtual member functions starting with the prefix *ml_*. The most important ones are the three virtual member functions of the form *ml_populate_*_feature_vector_()*, which construct the feature vectors associated to the node edit operations. Derived classes of *ged::MLBasedMethod* do not have to implement the machine learning frameworks, as *ged::MLBasedMethod* offers support for artificial deep neural networks (using FANN [24]) and support vector machines (using LIBSVM [13]). The following algorithms implemented in GEDLIB are directly derived classes of *ged::MLBasedMethod*: ged::BipartiteML [28], ged::RingML [4].

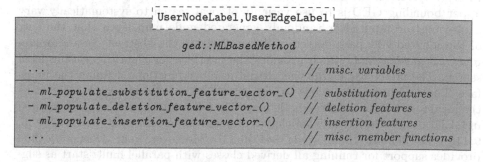

Fig. 5. The interface *ged::MLBasedMethod* for LSAPE based methods that use machine learning techniques for populating their LSAPE instances.

Interface for Methods Based on Mixed Integer Programming. Another approach for exactly or approximately computing GED is to rephrase the problem of computing GED(G, H) as a mixed integer programming (MIP) problem. GED(G, H) can then be computed exactly by calling an MIP solver. Alternatively, lower bounds for GED(G, H) can be obtained by solving the linear programming (LP) relaxations of the MIP formulations.

Figure 6 shows the abstract class template *ged::MIPBasedMethod*, which provides the interface for GED algorithms that use MIP formulations. The virtual member functions that define the interface start with the prefix *mip_*. The most important one is *mip_populate_model_()*, which constructs the employed MIP formulation and must be overridden by all derived classes. In GEDLIB, we use Gurobi [20] as our MIP and LP solver. Gurobi is commercial software but offers a free academic license. For users who cannot obtain a license for Gurobi, the installation script distributed with GEDLIB offers the option to install GEDLIB without *ged::MIPBasedMethod* and its derived classes. The following algorithms implemented in GEDLIB are directly derived classes of *ged::MIPBasedMethod*: ged::F1 [23], ged::F2 [23], ged::CompactMIP [6], ged::BLPNoEdgeLabels [21].

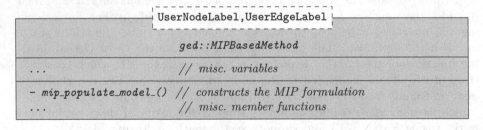

Fig. 6. The interface *ged::MIPBasedMethod* for methods based on MIP.

Interface for Methods Based on Local Search. Another popular approach for upper bounding GED is to use variants of local search to systematically vary a previously computed or randomly generated node map, such that the cost of the induced edit path decreases. Figure 7 shows the abstract class template *ged::LSBasedMethod*, which provides the interface for algorithms using local search. The prefix *ls_* marks the virtual member functions defining the interface. The most important one is *ls_run_from_initial_solution_()*, which runs the local search from an initial node map. The following algorithms implemented in GEDLIB are directly derived classes of *ged::LSBasedMethod*: ged::IPFP [5, 9,11], ged::BPBeam [16,29], ged::Refine [31]. Moreover, *ged::LSBasedMethod* provides support for running all derived classes with parallel multi-start as suggested in [15], and stochastic generators as suggested in [7].

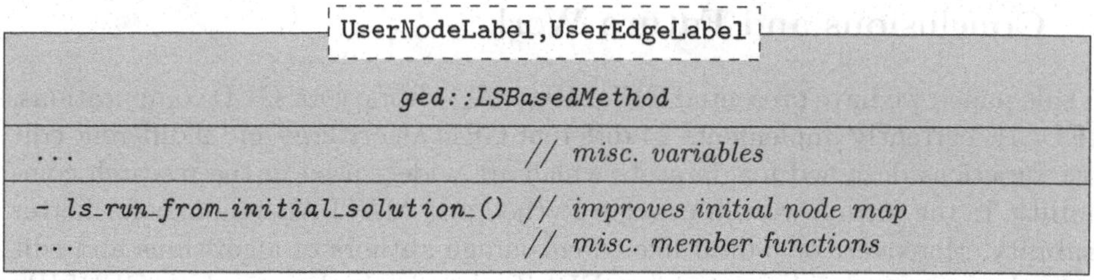

```
┌─────────────────────────────────────┐
│   UserNodeLabel,UserEdgeLabel        │
├─────────────────────────────────────┘─────┐
│              ged::LSBasedMethod            │
├────────────────────────────────────────────┤
│ ...                    // misc. variables  │
├────────────────────────────────────────────┤
│ - ls_run_from_initial_solution_()  // improves initial node map │
│ ...                    // misc. member functions │
└────────────────────────────────────────────┘
```

Fig. 7. The interface *ged::LSBasedMethod* for methods based on local search.

5   Abstract Class for Implementing Edit Costs

Figure 8 shows the abstract class template *ged::EditCosts*, which provided the interface for implementing edit cost functions. The virtual member functions **_del_cost_fun()* compute the cost of deleting a node or an edge with a given label, the functions **_ins_cost_fun()* compute the insertions costs, and the functions **_rel_cost_fun()* compute the costs for relabeling a node or an edge. The functions *vectorize_*_label()* return vector representations of the node and the edge labels, which are required by some methods. In GEDLIB, edit costs are available for the datasets AIDS, FINGERPRINT, GREC, LETTER, MUTA-GENICITY, and PROTEIN from the IAM Graph Database [25], for the datasets ACYCLIC, ALKANE, PAH, and MAO from GREYC's Chemistry Dataset (available at https://brunl01.users.greyc.fr/CHEMISTRY/), and for the dataset CMU-GED from the Graph Data Repository for Graph Edit Distance [1]. We also provide constant edit cost functions that can be used with any data.

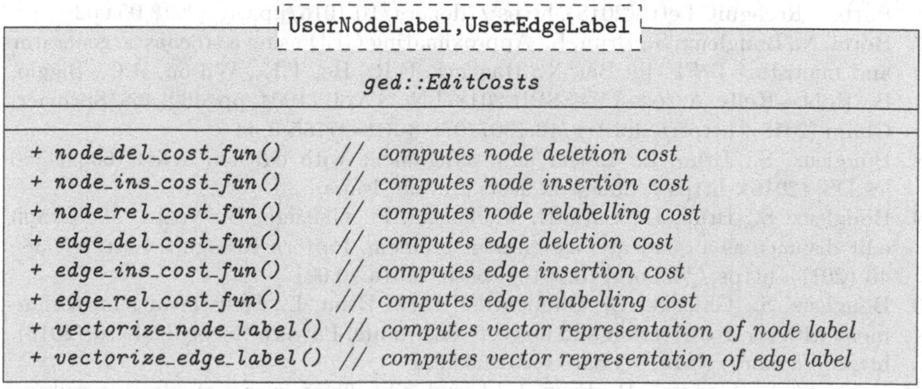

```
┌─────────────────────────────────────┐
│   UserNodeLabel,UserEdgeLabel        │
├─────────────────────────────────────┘─────────┐
│                  ged::EditCosts                │
├─────────────────────────────────────────────────┤
│ + node_del_cost_fun()   // computes node deletion cost      │
│ + node_ins_cost_fun()   // computes node insertion cost     │
│ + node_rel_cost_fun()   // computes node relabelling cost   │
│ + edge_del_cost_fun()   // computes edge deletion cost      │
│ + edge_ins_cost_fun()   // computes edge insertion cost     │
│ + edge_rel_cost_fun()   // computes edge relabelling cost   │
│ + vectorize_node_label() // computes vector representation of node label │
│ + vectorize_edge_label() // computes vector representation of edge label │
└─────────────────────────────────────────────────┘
```

Fig. 8. The interface *ged::EditCosts* for implementing edit costs.

6   Conclusions and Future Work

In this paper, we have presented GEDLIB, a C++ library for GED computations. GEDLIB currently implements 24 different GED algorithms and 9 different edit cost functions designed for datasets which are widely used in the research community. In the future, we will provide Python and MATLAB bindings for better usability. Moreover, we would like to encourage authors of algorithms and edit costs that are not implemented in GEDLIB to commit their work to GEDLIB.

References

1. Abu-Aisheh, Z., Raveaux, R., Ramel, J.-Y.: A graph database repository and performance evaluation metrics for graph edit distance. In: Liu, C.-L., Luo, B., Kropatsch, W.G., Cheng, J. (eds.) GbRPR 2015. LNCS, vol. 9069, pp. 138–147. Springer, Cham (2015). https://doi.org/10.1007/978-3-319-18224-7_14. http://www.rfai.li.univ-tours.fr/PublicData/GDR4GED/home.html
2. Blumenthal, D.B., Gamper, J.: Improved lower bounds for graph edit distance. IEEE Trans. Knowl. Data Eng. **30**(3), 503–516 (2018). https://doi.org/10.1109/TKDE.2017.2772243
3. Blumenthal, D.B., Bougleux, S., Gamper, J., Brun, L.: Ring based approximation of graph edit distance. In: Bai, X., Hancock, E.R., Ho, T.K., Wilson, R.C., Biggio, B., Robles-Kelly, A. (eds.) S+SSPR 2018. LNCS, vol. 11004, pp. 293–303. Springer, Cham (2018). https://doi.org/10.1007/978-3-319-97785-0_28
4. Blumenthal, D.B., Bougleux, S., Gamper, J., Brun, L.: Upper bounding GED via transformations to LSAPE based on rings and machine learning (2018, submitted)
5. Blumenthal, D.B., Daller, E., Bougleux, S., Brun, L., Gamper, J.: Quasimetric graph edit distance as a compact quadratic assignment problem. In: ICPR 2018, pp. 934–939 (2018)
6. Blumenthal, D.B., Gamper, J.: On the exact computation of the graph edit distance. Pattern Recognit. Lett. (2018). https://doi.org/10.1016/j.patrec.2018.05.002
7. Boria, N., Bougleux, S., Brun, L.: Approximating GED using a stochastic generator and multistart IPFP. In: Bai, X., Hancock, E.R., Ho, T.K., Wilson, R.C., Biggio, B., Robles-Kelly, A. (eds.) S+SSPR 2018. LNCS, vol. 11004, pp. 460–469. Springer, Cham (2018). https://doi.org/10.1007/978-3-319-97785-0_44
8. Bougleux, S., Brun, L.: Linear sum assignment with edition. arXiv:1603.04380 [cs.DS] (2016). https://bougleux.users.greyc.fr/lsape/
9. Bougleux, S., Brun, L., Carletti, V., Foggia, P., Gaüzère, B., Vento, M.: Graph edit distance as a quadratic assignment problem. Pattern Recognit. Lett. **87**, 38–46 (2017). https://doi.org/10.1016/j.patrec.2016.10.001
10. Bougleux, S., Gaüzère, B., Blumenthal, D.B., Brun, L.: Fast linear sum assignment with error-correction and no cost constraints. Pattern Recognit. Lett. (2018). https://doi.org/10.1016/j.patrec.2018.03.032
11. Bougleux, S., Gaüzère, B., Brun, L.: Graph edit distance as a quadratic program. In: ICPR 2016, pp. 1701–1706 (2016). https://doi.org/10.1109/ICPR.2016.7899881
12. Carletti, V., Gaüzère, B., Brun, L., Vento, M.: Approximate graph edit distance computation combining bipartite matching and exact neighborhood substructure distance. In: Liu, C.-L., Luo, B., Kropatsch, W.G., Cheng, J. (eds.) GbRPR 2015. LNCS, vol. 9069, pp. 188–197. Springer, Cham (2015). https://doi.org/10.1007/978-3-319-18224-7_19

13. Chang, C.C., Lin, C.J.: LIBSVM: a library for support vector machines. ACM Trans. Intell. Syst. Technol. **2**(3), 27 (2011). https://doi.org/10.1145/1961189. 1961199. https://www.csie.ntu.edu.tw/~cjlin/libsvm/

14. Chang, L., Feng, X., Lin, X., Qin, L., Zhang, W.: Efficient graph edit distance computation and verification via anchor-aware lower bound estimation. arXiv:1709.06810 [cs.DB] (2017)

15. Daller, É., Bougleux, S., Gaüzère, B., Brun, L.: Approximate graph edit distance by several local searches in parallel. In: ICPRAM 2018, pp. 149–158 (2018). https://doi.org/10.5220/0006599901490158

16. Ferrer, M., Serratosa, F., Riesen, K.: A first step towards exact graph edit distance using bipartite graph matching. In: Liu, C.-L., Luo, B., Kropatsch, W.G., Cheng, J. (eds.) GbRPR 2015. LNCS, vol. 9069, pp. 77–86. Springer, Cham (2015). https://doi.org/10.1007/978-3-319-18224-7_8

17. Fischer, A., Suen, C.Y., Frinken, V., Riesen, K., Bunke, H.: Approximation of graph edit distance based on Hausdorff matching. Pattern Recognit. **48**(2), 331–343 (2015). https://doi.org/10.1016/j.patcog.2014.07.015

18. Gaüzère, B., Bougleux, S., Riesen, K., Brun, L.: Approximate graph edit distance guided by bipartite matching of bags of walks. In: Fränti, P., Brown, G., Loog, M., Escolano, F., Pelillo, M. (eds.) S+SSPR 2014. LNCS, vol. 8621, pp. 73–82. Springer, Heidelberg (2014). https://doi.org/10.1007/978-3-662-44415-3_8

19. Guennebaud, G., Jacob, B., et al.: Eigen v3 (2010). http://eigen.tuxfamily.org

20. Gurobi Optimization, LLC: Gurobi optimizer reference manual (2018). http://www.gurobi.com

21. Justice, D., Hero, A.: A binary linear programming formulation of the graph edit distance. IEEE Trans. Pattern Anal. Mach. Intell. **28**(8), 1200–1214 (2006). https://doi.org/10.1109/TPAMI.2006.152

22. Lee, L., Lumsdaine, A., Siek, J.: The Boost Graph Library: User Guide and Reference Manual (2002). https://www.boost.org/doc/libs/1_68_0/libs/graph/doc/index.html

23. Lerouge, J., Abu-Aisheh, Z., Raveaux, R., Héroux, P., Adam, S.: New binary linear programming formulation to compute the graph edit distance. Pattern Recognit. **72**, 254–265 (2017). https://doi.org/10.1016/j.patcog.2017.07.029

24. Nissen, S.: Implementation of a fast artificial neural network library (FANN). Technical report, Department of Computer Science, University of Copenhagen (DIKU) (2003). http://leenissen.dk/fann/wp/

25. Riesen, K., Bunke, H.: IAM graph database repository for graph based pattern recognition and machine learning. S+SSPR 2008. LNCS, vol. 5342, pp. 287–297. Springer, Heidelberg (2008). https://doi.org/10.1007/978-3-540-89689-0_33. http://www.fki.inf.unibe.ch/databases/iam-graph-database

26. Riesen, K., Bunke, H.: Approximate graph edit distance computation by means of bipartite graph matching. Image Vis. Comput. **27**(7), 950–959 (2009). https://doi.org/10.1016/j.imavis.2008.04.004

27. Riesen, K., Bunke, H., Fischer, A.: Improving graph edit distance approximation by centrality measures. ICPR 2014, pp. 3910–3914 (2014). https://doi.org/10.1109/ICPR.2014.671

28. Riesen, K., Ferrer, M.: Predicting the correctness of node assignments in bipartite graph matching. Pattern Recognit. Lett. **69**, 8–14 (2016). https://doi.org/10.1016/j.patrec.2015.10.007

29. Riesen, K., Fischer, A., Bunke, H.: Combining bipartite graph matching and beam search for graph edit distance approximation. In: El Gayar, N., Schwenker, F., Suen, C. (eds.) ANNPR 2014. LNCS (LNAI), vol. 8774, pp. 117–128. Springer, Cham (2014). https://doi.org/10.1007/978-3-319-11656-3_11
30. Riesen, K., Fischer, A., Bunke, H.: Improved graph edit distance approximation with simulated annealing. In: Foggia, P., Liu, C.-L., Vento, M. (eds.) GbRPR 2017. LNCS, vol. 10310, pp. 222–231. Springer, Cham (2017). https://doi.org/10.1007/978-3-319-58961-9_20
31. Zeng, Z., Tung, A.K.H., Wang, J., Feng, J., Zhou, L.: Comparing stars: on approximating graph edit distance. PVLDB **2**(1), 25–36 (2009). https://doi.org/10.14778/1687627.1687631
32. Zheng, W., Zou, L., Lian, X., Wang, D., Zhao, D.: Efficient graph similarity search over large graph databases. IEEE Trans. Knowl. Data Eng. **27**(4), 964–978 (2015). https://doi.org/10.1109/TKDE.2014.2349924

Learning the Graph Edit Costs: What Do We Want to Optimise?

Elena Rica[(✉)], Susana Álvarez, and Francesc Serratosa

Universitat Rovira i Virgili, Tarragona, Catalonia, Spain
{mariaelena.rica, susana.alvarez,
francesc.serratosa}@urv.cat

Abstract. Graph edit distance has become an important tool in structural pattern recognition since it allows us to measure the dissimilarity of attributed graphs. One of its main constraints is that it requires an adequate definition of edit costs, which are application dependent. These costs eventually determine which graphs are considered similar or not in a concrete application. Several methods have been presented to learn these costs to avoid manually setting them. They are based on different techniques ranging from probabilistic methods to neural networks or known optimisation algorithms. The aim of this paper is twofold. On the one hand, we list them and summarize their features. On the other hand, we empirically analyse the behaviour of the proposed optimisation functions. We conclude that these functions return different edit costs and therefore, they have to be considered application dependent and not only a technicality of the method, as it has been considered so far.

Keywords: Graph edit distance · Learning edit costs · Optimisation function

1 Introduction

Graph edit distance [1–5] is one of the most well-known and used distance between attributed graphs. It is defined as the minimum amount of required distortion that transforms one graph into another. To this end, a number of distortion or edit operations consisting of deletion, insertion, and substitution of nodes and edges are defined. The basic idea is to assign an edit cost to each edit operation according to the amount of distortion that it introduces in the transformation; this allows to quantitatively evaluate the edit operations.

However, the structural and semantic dissimilarity of graphs is only correctly reflected by graph edit distance if the underlying edit costs are defined appropriately. For this reason, several methods have been presented to learn these edit costs. Nevertheless, the main features and differences of these methods have only been commented in the introduction of some papers but have never been summarised and put together in only one table.

The aim of this paper is twofold. On the one hand, we present a table showing the main features of eleven different methods that have been presented. On the other hand, we show, through a simple example, that the different proposed optimisation functions make the learning algorithm to return different edit costs.

D. Conte et al. (Eds.): GbRPR 2019, LNCS 11510, pp. 25–34, 2019.
https://doi.org/10.1007/978-3-030-20081-7_3

Until now, the scientific community considered that the edit costs were application dependent and therefore, they would have to be learned through examples generated by the same application (or database) that afterwards is going to be applied in pattern recognition. In addition, the optimisation function has to be considered application dependent.

This paper is structured as follows; Sect. 2 defines the attributed graphs and the graph edit distance. Section 3 summarises ten learning methods and their optimisation functions. Section 4 shows three optimisation functions given different graph edit distance parameters. Finally, Sect. 5 concludes the paper.

2   Attributed Graphs and Graph Edit Distance

Suppose we have a pair of graphs, G and G'. Also suppose the ith node in G is represented as G_i and the ath node in G' is represented as G'_a. Similarly, the edge between the G_i node and the G_j node in G is represented as $G_{i,j}$. And finally, the edge between G'_a node and the G'_b node in G' is represented as $G'_{a,b}$.

The graph edit distance between two attributed graphs consists in finding the best combination of edit operations that transforms one graph into another. Three operations are considered on the nodes and on the edges: Substitution, deletion and insertion. To quantify the distortion of each edit operation, a cost is assigned to them depending on the attributes on the involved nodes or edges. $C_s^n(i, a)$ is the cost of substituting the node G_i by the node G'_a, $C_D^n(i)$ is the cost of deleting the node G_i and $C_I^n(a)$ is the cost of inserting the node G'_a. Similarly, $C_s^e(i, a, j, b)$ is the cost of substituting the edge $G_{i,j}$ by the edge $G'_{a,b}$, $C_D^e(i,j)$ is the cost of deleting the edge $G_{i,j}$ and $C_I^e(a, b)$ is the cost of inserting the edge $G'_{a,b}$. Thus, the graph edit distance is defined as the transformation from one graph into another, through the edit operations, that obtains the minimum cost.

This graph transformation can be defined through a node-to-node mapping f between nodes of both graphs. In this way, we represent the mapping from node G_i to node G'_a as $a = f(i)$. We suppose both graphs have the same number of nodes since they have been expanded with new nodes that have a concrete attribute. We call these new nodes as *Null*. Note, the mapping between edges is imposed by the mapping of the nodes that these edges connect. Given the mapping $a = f(i)$ from node G_i to node G'_a, we say that represents a node substitution if both nodes are not *Null*. Contrarily, if node G'_a is a *Null* and G_i is not, we say that it represents a deletion. Finally, if node G_i is a *Null* and G'_a is not, we say that it represents an insertion. Similarly happens with the edges. The case that either nodes or both edges are *Null* is not considered since it is always defined with zero cost.

3 Learning Methods and Objective Functions

Table 1 shows the published methods related on learning the edit costs. There are ten papers from 2005 to 2018. The objective function is the most general term for any function to be optimized during training. The aim of the learning algorithm is to deduce the parameters of this function such this function reaches the global minimum. In some cases, the objective function is composed of the cost function plus a loss function. The cost function is the real aim of the optimisation process and the loss function is added to the objective function to have some control on the parameters to be learned.

Thus, Eq. (1) shows the general learning paradigm given the objective function of these methods where the parameters to be learned are the edit costs on nodes and edges,

$$
\begin{aligned}
&\left(\dot{C}_s^n, \dot{C}_D^n, \dot{C}_I^n, \dot{C}_s^e, \dot{C}_D^e, \dot{C}_I^e\right) \\
&= \max_{C_s^n, C_D^n, C_I^n, C_s^e, C_D^e, C_I^e} Objective\_function\left(C_s^n, C_D^n, C_I^n, C_s^e, C_D^e, C_I^e\right)
\end{aligned} \tag{1}
$$

The three most used objective functions for graph edit distance learning are the Dunn index [22], the recognition ratio and the correspondence accuracy.

In the first two of these objective functions, it is assumed that there is a database composed of classified graphs (a set of graphs that an oracle has imposed they belong to a class). Roughly speaking, the Dunn index is the ratio between the minimum of the distances between all the graphs that are classified at the same class and the maximum of the distances between all the graphs that are classified as having different classes. That is, it represents how much the classes are separated, considering the learned distance. The classification ratio informs of the number of correctly classified graphs normalised by the number of graphs in the test set.

In the correspondence accuracy case, the database of graphs is defined in a different way. Each register in the database is not composed of a graph and its class but a pair of graphs and their node-to-node correspondence. This correspondence has been imposed by an oracle. Then, for learning purposes, the classes of the graphs are not used. In [29], they present some public databases having this type of registers. The correspondence accuracy is the number of times a node-to-node mappings is the same than the one imposed by the oracle normalised by the number of node-to-node mappings.

Considering the edit costs, learning methods can be classified in two classes. The first class is composed of the methods that learn a function that represents the edit cost through a neural network, self organizing map or probability density representation. Thus, in the pattern recognition method, the edit costs are the output of these machine learning methods when the input are the attributes on the nodes or edges. In this case, the aim of the learning algorithm is to learn these functions. In Table 1, these edit costs are represented as $F_s^n, F_D^n, F_I^n, F_s^e, F_D^e$ and F_I^e.

The second class is composed of the methods that learn some constants. Then, in the pattern recognition process, the edit costs are computed through these constants and the machine learning method is not considered any more. In Table 1, these constants are represented as K_D^n, K_I^n, K_D^e and K_I^e. Moreover, the substitution costs are represented as a weighted Euclidean distance between the attributes on nodes of both graphs or between the edges of both graphs. In this case, the machine learning method learns these weights, which are represented as w_s^n and w_s^e.

Table 1. Published methods related on learning the graph edit costs

Ref.	Authors	Objective function	Learning method
2005 [6]	Neuhaus Bunke	Average of 8 indices: Davies–Bouldin [21] Dunn [22] C [23] Goodman–Krusk [24] Calinski–Haraba [25] Rand [26] Jaccard [27] Fowlkes–Mallo [28]	The method learns the weights of a Self Organized Map (SOM) to define the substitution, deletion and insertion costs on nodes and edges. These costs become the output of the SOM when the input is the attribute of the node or the edge **Learns**: $F_s^n, F_D^n, F_I^n, F_s^e, F_D^e, F_I^e$
2007 [7]	Neuhaus Bunke	Dunn Index [22]	The method learns the parameters of a Probability Density Function (PDF) to define the substitution, deletion and insertion costs on nodes and edges. These costs become the inverse of the probability set by the PDF given the attributes of the node or the edge **Learns**: $F_s^n, F_D^n, F_I^n, F_s^e, F_D^e, F_I^e$
2009 [8]	Caetano McAuley Cheng Le Smola	Correspondence accuracy	The method learns the weights of the weighted Euclidean distance to define the substitution cost on nodes and edges. The substitution cost becomes the weighted Euclidean distance for nodes and edges. Insertion and deletion of nodes and edges are not learned and assumed to be constant **Learns**: w_s^n, w_s^e
2012 [9]	Leordeanu Sukthankar Hebert	Recognition ratio	The same than [8]
2015 [10]	Cortés Serratosa	Correspondence accuracy	The method learns the deletion and insertion costs on nodes and edges as constants (Real numbers). The substitution cost is assumed the Euclidean distance between the attributes on nodes or on edges **Learns**: $K_D^n, K_I^n, K_D^e, K_I^e$
2016 [11]	Cortés Serratosa	Correspondence accuracy	The same than [8]
2017 [12]	Raveaux Martineau Conte Venturini	Recognition ratio	The same than [8]

(continued)

Table 1. (*continued*)

Ref.	Authors	Objective function	Learning method
2018 [13]	Cortés Conte Cardot Serratosa	Correspondence accuracy	The method learns the substitution functions on nodes and edges through a Neural Network (NN). The substitution cost is defined as the output of the NN when the input is the attribute on the nodes and edges. Insertion and deletion of nodes and edges are not learned and assumed to be constants **Learns**: F_s^n, F_s^e
2018 [14]	Santacruz Serratosa	Correspondence accuracy	Similar to [13] but the insertion and deletion costs on nodes and edges are also learned. There is also a NN for insertion and another one for deletion the nodes and edges. **Learns**: $F_s^n, F_D^n, F_I^n, F_s^e, F_D^e, F_I^e$
2018 [15]	Algabli Serratosa	Correspondence accuracy	The method learns the weights of the weighted Euclidean distance to define the substitution cost and also the deletion and insertion costs as constants on nodes and edges. The substitution cost is computed as a weighted Euclidean distance in which the weights have been learned. The insertion and deletion costs become the learned constant (Real number) **Learns**: $w_s^n, K_D^n, K_I^n, w_s^e, K_D^e, K_I^e$
2018 [16]	Martineau Raveaux Conte Venturini	Recognition ratio	The method learns the weights on each node or edge. These weights depend on how important are the nodes and edges to describe de class **Learns**: weights on nodes and edges

Papers [17] and [18] published in 2008 and 2013, respectively, have not been included in the table since, although they are related to learning graph matching parameters, are image registration oriented. Therefore, they are not general graph matching learning algorithms. Similarly happens with [19, 20], which are dedicated to learning or deducing the optimal correspondence through human interaction

4 Experimental Evaluation

The aim of this experimental evaluation is to show that the optimisation functions based on the correspondence accuracy, the recognition ratio and the Dunn index maximise at different edit cost values. Thus, the decision of which optimisation function has to be used becomes application dependent and not only a technicity in the learning algorithm.

The experimental evaluation has been carried out using four different databases: Letter_High, Letter_Med, Letter_Low and House_Hotel [29]. The main characteristic of them is that their registers are not only composed of a graph and its class, but they are composed of a pair of graphs and a ground-truth correspondence between them, as well as their class. This register structure is useful to analyse and develop graph matching algorithms and to learn their parameters in a broad manner.

The Letter_High, Letter_Med and Letter_Low databases consists in a set of graphs that represent artificially distorted letters of the Latin alphabet. For each class, a prototype line drawing was manually constructed. These prototype drawings are then converted into prototype graphs by representing the lines through undirected edges, and the ending points of such lines through nodes. Attributes on nodes are the bi-dimensional position of the stroke junctions and edges do not have attributes. The three variants of the database depend on the degree of distortion with respect to the original prototype (adding, deleting and moving nodes and edges). The House_Hotel database consist of 111 graphs corresponding to a toy house and 101 graphs corresponding to a hotel. Each frame of these sequences has the same 30 hand-marked salient points identified and labelled with some attributes. Therefore, nodes in the graphs represent the salient points, with their position in the image plus a 60-size feature vector using Context Shape as attributes. Edges are imposed by the Delaunay triangulation.

Figures 1, 2, 3 and 4 show the correspondence accuracy, the classification ratio and the Dunn index of all the graphs in the test set. Moreover, they also show the mean of these three optimisation functions computed on the Letter_High, Letter_Med, Letter_Low and House_Hotel databases. In the four cases, we have only computed these functions given several values of the insertion and deletion costs on nodes and edges. The substitution costs have been computed as the Euclidean distance (all the weights on the attributes have the same value). The node insertion costs and the edge insertion costs have been set equal to node deletion costs and the edge deletion costs, respectively.

The first we realise is that most of the surfaces have several local maximum values, which makes difficult the optimisation algorithm to find the global maximum of the function. Moreover, these local maximum values and the global ones in the three optimisation functions (correspondence accuracy, the classification ratio and the Dunn index) do not appear in the same positions of the insertion and deletion values. Finally, the maximum of the mean of these functions (lower right plot) does not appear in any of the maximum of the three functions, since they are completely different. Nevertheless, in general, the mean seems to be smoother than the optimisation functions and therefore, usually easier to find the global maximum using a suboptimal algorithm. Note that the method in [6] is based on optimising the mean of eight optimisation algorithms and we assume it was done in this way due to this feature.

As previously commented, the difference between the three Letter databases is the increasing noise applied on the graphs. Thus, we were tempted to find a relation between the maximisation points and the amount of noise. For instance, when the noise increases, the optimal insertion and deletion costs also tend to increase. Nevertheless, we have not been able to find these behaviours.

Finally, all combinations of insertion and deletion costs in the analysed domain on nodes and edges in the House_Hotel database return a classification ratio of one. Nevertheless, the correspondence accuracy and the Dunn index are strongly dependent

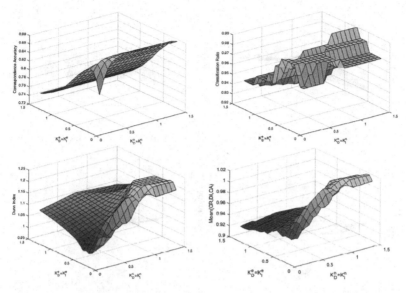

Fig. 1. Correspondence accuracy, classification ratio, Dunn index and the mean of these functions applied to the Letter_High database and given several combinations of insertion and deletion costs on nodes and edges.

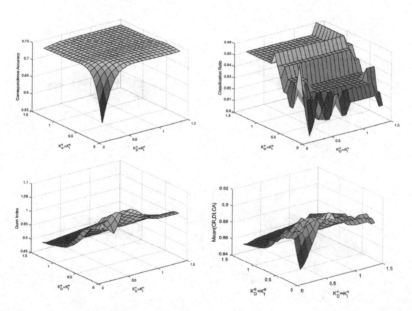

Fig. 2. Correspondence accuracy, classification ratio, Dunn index and the mean of these functions applied to the Letter_Med database and given several combinations of insertion and deletion costs on nodes and edges.

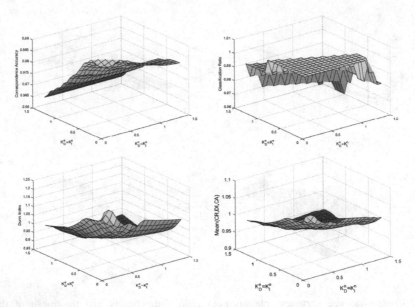

Fig. 3. Correspondence accuracy, classification ratio, Dunn index and the mean of these functions applied to the Letter_Low database and given several combinations of insertion and deletion costs on nodes and edges.

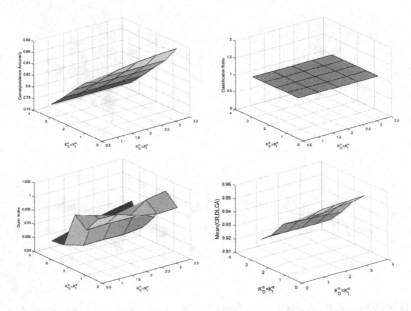

Fig. 4. Correspondence accuracy, classification ratio, Dunn index and the mean of these functions applied to the House_Hotel database and given several combinations of insertion and deletion costs on nodes and edges.

on these costs. Thus, in this case, optimising on the classification ratio seems not to have much sense. Thus, this other example also strengthens the idea that the optimisation function has to be considered application dependent.

5 The Conclusions

Edit costs are application dependent and frequently set manually. Nevertheless, some methods have been published to learn these edit costs. The aim of this paper is twofold. On the one hand, we have summarised ten of these methods in only one table and highlighted their differences. On the other hand, we have analysed the effect of the three optimisation functions proposed by these methods, which are the correspondence accuracy, the classification ratio and the Dunn index.

Given a simple experiment on four databases, we have concluded that the maximisation of the three optimisation functions return different edit costs and therefore, the use of one of these functions has to be deeply considered and it is clearly application dependent. Finally, in the case that we want to have a learning algorithm independent of the optimisation function, a useful option could be to optimise the mean of these three functions. This option has the advantage that the mean of the optimisation functions appears to be smoother than the other three functions but it has the drawback that the computational cost drastically increases.

Acknowledgments. This research is supported by projects TIN2016-77836-C2-1-R, DPI2016-78957-R AEI/FEDER EU; and the European projects AEROARMS, H2020-ICT-2014-1-644271 and NanoInformaTIX, H2020-NMBP-TO-IND-2018-2020-814426.

References

1. Bunke, H., Allermann, G.: Inexact graph matching for structural pattern recognition. Pattern Recognit. Lett. **1**(4), 245–253 (1983)
2. Sanfeliu, A., Fu, K.S.: A distance measure between attributed relational graphs for pattern recognition. IEEE Trans. Syst. Man Cybern. **13**(3), 353–362 (1983)
3. Gao, X., Xiao, B., Tao, D., Li, X.: A survey of graph edit distance. Pattern Anal. Appl. **13**(1), 113–129 (2010)
4. Serratosa, F., Cortés, X.: Graph Edit Distance: moving from global to local structure to solve the graph-matching problem. Pattern Recognit. Lett. **65**, 204–210 (2015)
5. Serratosa, F.: Graph edit distance: Restrictions to be a metric. Pattern Recognit. **90**, 250–256 (2019)
6. Neuhaus, M., Bunke, H.: Self-organizing maps for learning the edit costs in graph matching. Trans. Syst. Man Cybern. **35**(3), 305–314 (2005)
7. Neuhaus, M., Bunke, H.: Automatic learning of cost functions for graph edit distance. Inf. Sci. **177**(1), 239–247 (2007)
8. Caetano, T.S., McAuley, J.J., Cheng, L., Le, Q.V., Smola, A.J.: Learning graph matching. Trans. Pattern Anal. Mach. Intell. **31**(6), 1048–1058 (2009)
9. Leordeanu, M., Sukthankar, R., Hebert, M.: Unsupervised learning for graph matching. Int. J. Comput. Vis. **96**(1), 28–45 (2012)

10. Cortés, X., Serratosa, F.: Learning graph-matching edit-costs based on the optimality of the Oracle's node correspondences. Pattern Recognit. Lett. **56**, 22–29 (2015)
11. Cortés, X., Serratosa, F.: Learning graph matching substitution weights based on the ground truth node correspondence. Int. J. Pattern Recognit. Artif. Intell. **30**(2), 1650005, 22 p. (2016)
12. Raveaux, R., Martineau, M., Conte, D., Venturini, G.: Learning graph matching with a graph-based perceptron in a classification context. In: Foggia, P., Liu, C.-L., Vento, M. (eds.) GbRPR 2017. LNCS, vol. 10310, pp. 49–58. Springer, Cham (2017). https://doi.org/10. 1007/978-3-319-58961-9_5
13. Cortés, X., Conte, D., Cardot, H., Serratosa, F.: A deep neural network architecture to estimate node assignment costs for the graph edit distance. In: Bai, X., Hancock, E.R., Ho, T.K., Wilson, R.C., Biggio, B., Robles-Kelly, A. (eds.) S+SSPR 2018. LNCS, vol. 11004, pp. 326–336. Springer, Cham (2018). https://doi.org/10.1007/978-3-319-97785-0_31
14. Santacruz, P., Serratosa, F.: Learning the sub-optimal graph edit distance edit costs based on an embedded model. In: Bai, X., Hancock, E.R., Ho, T.K., Wilson, R.C., Biggio, B., Robles-Kelly, A. (eds.) S+SSPR 2018. LNCS, vol. 11004, pp. 282–292. Springer, Cham (2018). https://doi.org/10.1007/978-3-319-97785-0_27
15. Algabli, S., Serratosa, F.: Embedding the node-to-node mappings to learn the Graph edit distance parameters. Pattern Recognit. Lett. **112**, 353–360 (2018)
16. Martineau, M., Raveaux, R., Conte, D., Venturini, G.: Learning error-correcting graph matching with a multiclass neural network. Pattern Recognit. Lett. (2018, in press). https:// doi.org/10.1016/j.patrec.2018.03.031
17. Torresani, L., Kolmogorov, V., Rother, C.: Feature correspondence via graph matching: models and global optimization. In: Forsyth, D., Torr, P., Zisserman, A. (eds.) ECCV 2008. LNCS, vol. 5303, pp. 596–609. Springer, Heidelberg (2008). https://doi.org/10.1007/978-3-540-88688-4_44
18. Cho, M., Alahari, K., Ponce, J.: Learning graphs to match. In: ICCV 2013, pp: 25–32 (2013)
19. Serratosa, F., Cortés, X.: Interactive graph-matching using active query strategies. Pattern Recognit. **48**(4), 1364–1373 (2015)
20. Cortés, X., Serratosa, F.: An interactive method for the image alignment problem based on partially supervised correspondence. Expert Syst. Appl. **42**(1), 179–192 (2015)
21. Davies, D., Bouldin, D.: A cluster separation measure. IEEE Trans. Pattern Anal. Mach. Intell. **PAMI-1**(2), 224–227 (1979)
22. Dunn, J.: Well-separated clusters and optimal fuzzy partitions. J. Cybern. **4**, 95–104 (1974)
23. Hubert, L., Schultz, J.: Quadratic assignment as a general data analysis strategy. Br. J. Math. Stat. Psychol. **29**, 190–241 (1976)
24. Goodman, L., Kruskal, W.: Measures of Association for Cross Classification. Springer, New York (1979). https://doi.org/10.1007/978-1-4612-9995-0_1
25. Calinski, T., Harabasz, J.: A dendrite method for cluster analysis. Commun. Stat.-Theory Methods **3**(1), 1–27 (1974)
26. Rand, W.: Objective criteria for the evaluation of clustering methods. J. A. Stat. Assoc. **66**(336), 846–850 (1971)
27. Jain, A., Dubes, R.: Algorithms for Clustering Data. Prentice-Hall, Englewood Cliffs (1988)
28. Fowlkes, E., Mallows, C.: A method for comparing two hierarchical clusterings. J. Am. Stat. Assoc. **78**, 553–584 (1983)
29. Moreno-García, C.F., Cortés, X., Serratosa, F.: A graph repository for learning error-tolerant graph matching. In: Robles-Kelly, A., Loog, M., Biggio, B., Escolano, F., Wilson, R. (eds.) Structural, Syntactic, and Statistical Pattern Recognition, vol. 10029, pp. 519–529. Springer, Cham (2016). https://doi.org/10.1007/978-3-319-49055-7_46

Sub-optimal Graph Matching
by Node-to-Node Assignment Classification

Xavier Cortés[1(✉)], Donatello Conte[1], and Francesc Serratosa[2]

[1] Laboratoire d'Informatique Fondamentale et Appliquée de Tours
(LIFAT - EA6300), Tours, France
{xavier.cortes,donatello.conte}@univ-tours.fr
[2] Universitat Rovira i Virgili, Tarragona, Catalonia, Spain
francesc.serratosa@urv.cat

Abstract. In the recent years, Graph Edit Distance has awaken interest in the scientific community and some new graph-matching algorithms
that compute it have been presented. Nevertheless, these algorithms usually cannot be used in real applications due to runtime restrictions.
For this reason, other graph-matching algorithms have also been used
that compute an approximation of the graph correspondence with lower
runtime. Clearly, in a real application, there is a tradeoff between runtime and accuracy. One of the most costly part in these algorithms is
the deduction of the node-to-node mapping. We present a new graph-
matching algorithm that returns a graph correspondence without the
explicit computation of the assignment problem. This is done thanks
to a classification of the node-to-node assignment learned in a previous
training stage.

Keywords: Graph Edit Distance · Node assignment classification ·
Graph embedding · Graph matching

1 Introduction

Attributed graphs have found widespread applications in several research fields
of structural pattern recognition [1–3]. This is due to their ability to represent
structured objects through unary and binary local entities. To compare them,
several distance measures between attributed graphs have been presented [2,3].
Among them, one of the most used distances is the Graph Edit Distance [4,5].

Typically, the problem is mathematically formulated as a quadratic assignment
problem, which consists of finding the node-to-node assignment that minimizes an
objective function encoding local dissimilarities (a linear term) and structural dissimilarities (a quadratic term). To do so, it is needed to define the cost functions
between the linear terms and also the quadratic terms, given the application at
hand. Note that a proper definition of these cost functions is crucial to achieve good
classification or recognition results. For this reason, several methods have been presented to learn these edit costs [6,7]. Moreover, the Graph Edit Distance has been

© Springer Nature Switzerland AG 2019
D. Conte et al. (Eds.): GbRPR 2019, LNCS 11510, pp. 35–44, 2019.
https://doi.org/10.1007/978-3-030-20081-7_4

demonstrated to be an NP-hard problem. For this reason, several algorithms that return a graph correspondence in polynomial time with respect to the number of nodes at the expense of not having the certainty of being the correspondence that minimizes the graph edit distance, have been presented [8–12].

In this paper, we present another graph matching algorithm that deduces the graph correspondences in a sub-optimal way, as the ones commented above. The novelty of our algorithm is that we classify the node-to-node assignments at a first step and then there is no need of using a combinatorial optimization algorithm (such as the Hungarian method [13]) to deduce the graph correspondence, thus avoiding their computational cost.

2   Definitions

2.1   Attributed Graphs and Graph Edit Distance

We define an attributed graph G as a quadruplet $G = \{\sum_v, \sum_e, \gamma_v, \gamma_e\}$ where $\sum_v = \{v_i | i = 1...n\}$ is the set of n nodes and $\sum_e = \{e_{i,j} | i, j \in 1...n\}$ is the set of edges connecting pairs of nodes. $\gamma_v = \{v_i \rightarrow \psi_i | i = 1..n\}$ and $\gamma_e = \{v_i \rightarrow E(v_i) | i = 1..n\}$ are functions that map the nodes and edges to their attribute values, respectively. $\psi_i \in \mathbb{R}^m$ maps each node to its m local attributes and $E(\cdot)$ refers to the degree of a certain node [14, 15]. For simplicity, in this paper we only consider undirected and unattributed edges. However, all the concepts presented in this paper could be extended to directed and attributed edges.

The Graph Edit Distance (GED) [4] is a distance between two attributed graphs G and G'. It consists of the best combination of edit operations that transform G into G'. Three operations are considered on the local attributes of the nodes and also on its structures: Substitution, deletion and insertion. To quantify the degree of distortion that each edit operation introduces, a cost is assigned to them depending on the attributes on the involved nodes or edges. A sequence of edit operations that completely transform G into G' is referred to as edit path $\lambda_{G,G'}$ between G and G'. The cost of an edit path is the sum of the costs of the edit operations included on it. Thus, the GED is defined as the edit path from one graph into another that obtains the minimum cost under all possible edit paths $\mathcal{T}_{G,G'}$ between G and G'. Formally:

$$GED(G, G') = \min_{\lambda_{G,G'} \in \mathcal{T}_{G,G'}} Cost(G, G', \lambda_{G,G'}) \qquad (1)$$

The edit path $\lambda_{G,G'}$ can be defined through a node-to-node matching f between nodes of both graphs where $f(v_i) = v'_a$. Graphs can be enlarged by null nodes to assure having the same order.

2.2   Approximating the Graph Edit Distance

The graph-matching algorithms that return the GED are based on exploring all the combinations of correspondences f between G and G' and selecting the one with the minimum cost. However, several approaches have appeared to reduce the computational complexity of the GED computation [9,16–18] at the expense of returning a sub-optimal correspondence. Usually, these algorithms are based on two main steps:

In the first step, a cost matrix is filled with the edit cost between all combinations of the local structures of both graphs. The computational cost of this step is approximated by $O(s \times n^2)$, where s is the computational cost of mapping the local structures.

In the second step, a node-to-node matching f is found. The problem at hand is seen as a minimization of the sum of the linear assignation given the cost matrix. Thus, the computational cost of this second step is $O(n^3)$ or $O(n^2)$ depending on whether the matching is deduced by the Bipartite graph matching [9,18], or the Greedy edit matching [8,14,16,17], respectively. In the case of the Bipartite graph matching, it is usually solved through the Munkres or Jonker-Volgenant algorithms [13,19]. In another case, the matching between nodes f is obtained through an algorithm that iteratively selects the minimum value per row and discards the selected columns for the remaining rows.

3   Learning Graph Matching

We propose a sub-optimal graph matching algorithm which avoids the second step in the classical sub-optimal graph matching algorithms (see Sect. 2.2). This is because the second part turns out to be more expensive than the first part, from the computational time point of view. In the next section, we list the known methods that learn the edit costs, from which our method is inspired.

3.1   Learning the Edit Costs and Graph Embedding

Several methods have been presented to learn the edit costs based on supervised machine learning techniques, which can be divided in two main groups. The ones that return a constant on the edit operations [20–24], and the other ones that define the edit costs as functions. For instance, in [25], they use a probabilistic model of the distribution of graph edit operations. Another paper is based on a self-organizing map model [26] in which the edit costs are the output of a Neural Network. In both papers, the learning set is composed of classified graphs and the edit costs are optimized with regard to Dunn's index [27]. Recently, two new papers assume the cost matrix could filled as the output of a supervised machine learning model. In [7], the authors use a Neural Network to learn only the substitution costs (no insertion nor deletion operations are allowed). And in [6], a general framework is presented to learn and define this costs.

3.2   From Edit Costs Estimation to Node Assignment Classification

Inspired by methods such as the ones in [6,7], we propose a supervised machine learning model that splits the node-to-node assignments in two classes, depending whether the learning database considers that they have to be mapped in f

or not. Note that in [6,7], the learning algorithms deduce edit costs instead of discerning between two classes. The key idea of our model is to decide if a node in G is mapped to a node in G' using a classifier. Then, the classes *mapped* or *non-mapped* are assigned considering the output of a previously trained classifier. Note that a node could remain unmapped in f if it is classified as *non-mapped* in all cases. We treat this particular case as a deletion or an insertion of the corresponding node.

Our method is independent of the classification method (Support Vector Machine, K-Nearest Neighbours, Neural Network, etc.), however, in any case we need to transform each node-to-node mapping into a vector that becomes the input of the classifier. Thus, we propose to embed each matching into a vector, similar as proposed in [7]. In this case, the embedded representation of a mapping between two nodes v_i and v'_a of G and G', is $x_{v_i \to v'_a} = [\gamma_v(v_i), \gamma_e(v_i), \gamma'_v(v'_a), \gamma'_e(v'_a)] \in \mathbb{R}^{(m+1) \cdot 2}$.

Our matching algorithm is shown in Algorithm 1. It is a greedy algorithm that goes across all nodes $v_i \in G$ and, for each of them, deduces the first node $v'_a \in G'$ that can be mapped to v_i. Note that this strategy avoids: (a) The explicit computation of the graph correspondences in the second step of the classical sub-optimal graph matching algorithms. (b) The whole computation of the cost matrix in the first step of the classical sub-optimal graph matching algorithms. Note that in our case, instead of having a cost matrix, we have a classification matrix.

Nevertheless, it is important to remark that the model does not return a distance but only a graph correspondence. Moreover, the performance of the model depends on the quality of the classifier.

Algorithm 1. Graph matching based on node assignment classification.

Data: graph G and graph G'
Result: matching f
1 f = empty node-to-node correspondence
2 **forall the** v_i *in* G **do**
3 | **forall the** v'_a *in* G' *and not in* f **do**
4 | | $x = embed(v_i \to v'_a)$
5 | | $y = class\_predictor(x)$
6 | | **if** $y = mapped$ **then**
7 | | | $f(v_i) = v'_a$
8 | | | break_loop
9 | **end**
10 **end**
11 **end**

Figure 1 (on the left) shows a pair of graphs (blue and red) and the optimal graph correspondence (green arrows), edges do not have attributes and the node edit cost is the distance between attributes. In the center of Fig. 1 the cost matrix computed in the first step of a sub-optimal graph-matching algorithm (Sect. 2.2) is shown. Finally, Fig. 1 (on the right) shows the outputs of the classifier (T: *mapped* or F: *non-mapped*) by our proposed algorithm. Grey cells are the node pairs that our algorithm needs to analyze. In our example, only 7 values have been computed. Being 4 the minimum and 10 the maximum number in the worst case (n and $n(n+1)/2$, respectively, where n is the number of nodes).

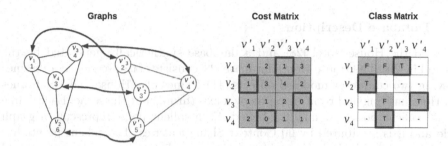

Fig. 1. Left: A pair of graphs (blue and red) and the optimal correspondence (green arrows). Centre: The cost matrix. Right: The computed classes (T: *mapped* or F: *non-mapped*). Highlighted in green: Node-to-node mappings. Grey: Computed values (nodes are processed consecutively from 1 to 4). (Color figure online)

3.3 Training the Classifier

In supervised machine learning, databases entries are composed of an element and its expected outputs. In our case, an entry p in the database is composed of a pair of graphs $(G^p, G^{p'})$ and their ground-truth correspondence $\hat{f}^p$. These correspondences $\hat{f}^p$ have been deduced by an external system (typically a human expert) and they are considered to be the best mappings for our learning purposes. The aim of the learning method is to define a model that, given a pair of nodes, returns the class *mapped* when the two nodes are mapped in the ground-truth, and *non-mapped* otherwise.

To do so, we define two sets of node-to-node mappings: The one with node pairs that have to be mapped according the ground-truth correspondences and the set of node pairs assumed they do not have to be mapped. Assuming that the ground-truth correspondences are bijective functions, each node of G is mapped to a single node of G', while it has not to be mapped to the rest of nodes of G'. This means that for each node-to-node mapping classified as *mapped*, there are $n-1$ node-to-node mappings classified as *non-mapped*, where n is the order of the graphs. In order to prevent imbalance problem, the node-to-node mappings in *mapped* are repeated $n-1$ times when we feed the training algorithm.

4   Experimental Evaluation

In this section, we show the performance of our graph-matching algorithm in terms of accuracy and runtime. To do so, we compare our results to different existing approaches in the literature. In Sect. 4.1, we describe the databases used in our experiments while in Sect. 4.2, we present the computed accuracy and runtime. Finally, in Sect. 4.3, we evaluate the performance of our method using synthetically generated graphs to analyze how the graph order affects our model. Experiments were conducted using MATLAB 2018 on a Windows 10, with an Intel i5 processor at 1.6 GHz and 8 GB of RAM.

4.1   Database Description

We used the House-Hotel [28] and a database synthetically generated by the method in [29]. The House-Hotel database [28] consists of two sequences of frames showing two computer modeled objects, 111 frames of a House and 101 frames of a Hotel moving and rotating on its own axis throughout the scene. Each frame has 30 salient points manually labelled. Each salient point represents a graph node and it is attributed by 60 Context Shape features. The salient points has been triangulated by Delaunay triangulation according to its coordinates to build the edges. Since the salient points are manually labelled, we know the ground-truth correspondence between the nodes of the graphs. As more separated are the frames that represents each graph, the differences between graphs increase and consequently, more difficult is the graph matching process. We performed different experiments changing the number of frames of separation between the frames in the video sequence. For each experiment, we built three sub-sets of graph pairs (train, validation and test).

The synthetic database was composed of several sets of graphs that had the same order. We generated different pairs of graph prototypes inspired by the method detailed in [29]. Nodes has four Integer attributes randomly generated in a range from 0 to 999. Edges are unattributed and it was imposed a probability of 20% for each pair of nodes to be connected by an edge. Next, a collection of pairs of graphs has been generated by distorting original prototypes adding some Gaussian noise (Standard deviation: 500) to the last attribute of each node.

4.2   Graph Matching Performance

We analyzed the graph matching accuracy and the time spent to perform it. The matching accuracy is defined as the number of node-to-node mappings in the deducted correspondences that are equal to the node-to-node mappings in the ground-truth correspondences, normalized by the graph order.

Our method was implemented with two different classifiers: (a) A Neural Network with one hidden layer that has 60 neurons and an output layer with two neurons (a neuron per class: *mapping* and *non-mapping*). (b) A Quadratic Support Vector Machine, which classifies the node-to-node mappings among the two classes: *mapping* and *non-mapping*.

Moreover, we compared our approach to four graph-matching methods (Table 1). In the first two methods, the cost function is defined as an edit cost based on the Euclidean distance between the node attributes plus the difference of the degree of these attributes, as it was done in [14,15]. In the other two, the cost function is defined as the output of a Neural Network previously trained as in [7].

Table 1. Graph matching methods used in this paper for comparative purposes.

Cost function	Solver	Complexity	Year	Reference
Edit Cost	Hungarian	$O(n^3)$	2009	[9]
Edit Cost	Greedy	$O(n^2)$	2017	[16]
Neural Network	Hungarian	$O(n^3)$	2018	[7]
Neural Network	Greedy	$O(n^2)$	-	-

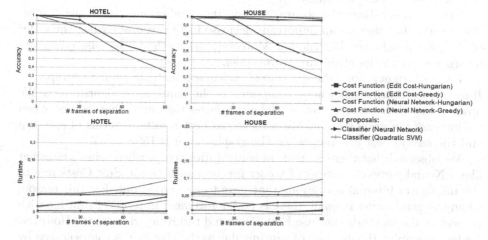

Fig. 2. Matching accuracy (top) and Runtime in seconds (bottom) versus number of frames of separation. Database: Hotel-House.

Figure 2 (top) shows the matching accuracy versus the number of frames of separation using the Hotel-House database. When the number of frames of separation increases, graphs tend to be more different and this fact is reflected on the accuracy decrease. On the bottom of the Fig. 2 the mean runtime spent to perform a matching between two graphs (in seconds) is shown. There is a slight tendency of increasing the runtime.

In these experiments, our method shows a good balance on runtime and matching accuracy. The accuracy is similar to the Neural Network methods [7] but the runtime is lower. Moreover, the GED methods [9,16] are faster but return a worse matching accuracy when the number of frames of separation increases.

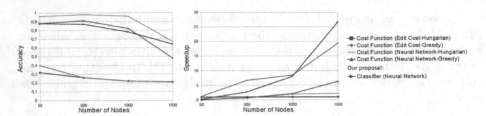

Fig. 3. Matching accuracy and speedup with respect to the GED-Hungarian versus number of nodes per graph. Database: Synthetic.

4.3   Runtime Analysis

We analyzed the relevance of the graph order with regard to the performance of the model in the synthetic database. Our model was the one implemented with a Neural Network that has 6 neurons in the hidden layer. The Quadratic Support Vector Machine was not used in this experiment due to the high computational time necessary in the learning step.

Figure 3 (left) shows the accuracy versus the number of nodes of the graphs. We observe that there is an important decrease in the accuracy when the order of the graphs achieves 1500 nodes. Moreover, the methods that use edit costs return low accuracies given any graph order.

Figure 3 (right) shows the speedup of each graph matching alg: $Speedup_{alg} = Runtime_{[9]}/Runtime_{alg}$, where $Runtime_x$ is the runtime of algorithm x. We observe that the speedup of our model grows faster than any other method when we increase the graphs order. We achieve the best results in terms of speedup and the second best accuracy when the graphs order is 1500 nodes.

We observed that there is an extra computational cost when using a classifier like a Neural Network instead of a cost function such as the Edit Costs due to the number of internal operations that have to be carried out. For this reason, when the graph order is small, there is no improvement in terms of runtime with respect to the methods that use Edit Costs and the Hungarian solver. However, for larger graphs, the increase of runtime due to the classifier is compensated by the fact that we do not need to evaluate all node-to-node correspondences and it is not necessary to solve the assignment problem either.

5   Conclusions

In this paper, we present a fast approach to deduce the graph correspondence. Previous methods are based on two steps. In the first one, a cost matrix is filled and in the second one, a linear solver is applied on it to deduce the graph correspondence. In our proposal, we do not need this second step since the first one, the node-to-node mapping, is directly deduced. To do so, we have used a classifier that separates the node-to-node assignment in two classes: *mapped* and *non-mapped*. In the experimental section, we show that our method achieves a good accuracy with a low matching runtime, comparing it to four existing methods. The experiments show a larger decrease of runtime, compared to the other methods, when the graph order increases. This allows to compute matchings between graphs with the proposed method for very large graphs in an acceptable computational time.

Acknowledgements. This research is supported by projects TIN2016-77836-C2-1-R and DPI2016-78957-R and AEROARMS (H2020-ICT-2014-1-644271 (Spain), and by DANIEAL2 project of Centre-Val-de-Loire Region (France).

References

1. Conte, D., Foggia, P., Sansone, C., Vento, M.: Thirty years of graph matching in pattern recognition. Int. J. Pattern Recognit. Artif. Intell. **18**(03), 265–298 (2004)
2. Livi, L., Rizzi, A.: The graph matching problem. Pattern Anal. Appl. **16**(3), 253–283 (2013)
3. Foggia, P., Percannella, G., Vento, M.: Graph matching and learning in pattern recognition in the last 10 years. Int. J. Pattern Recognit. Artif. Intell. **28**(01), 1450001 (2014)
4. Gao, X., Xiao, B., Tao, D., Li, X.: A survey of graph edit distance. Pattern Anal. Appl. **13**(1), 113–129 (2010)
5. Serratosa, F.: Graph edit distance: restrictions to be a metric. Pattern Recognit. **90**, 250–256 (2019)
6. Santacruz, P., Serratosa, F.: Learning the sub-optimal graph edit distance edit costs based on an embedded model. In: Bai, X., Hancock, E.R., Ho, T.K., Wilson, R.C., Biggio, B., Robles-Kelly, A. (eds.) S+SSPR 2018. LNCS, vol. 11004, pp. 282–292. Springer, Cham (2018). https://doi.org/10.1007/978-3-319-97785-0_27
7. Cortés, X., Conte, D., Cardot, H., Serratosa, F.: A deep neural network architecture to estimate node assignment costs for the graph edit distance. In: Bai, X., Hancock, E.R., Ho, T.K., Wilson, R.C., Biggio, B., Robles-Kelly, A. (eds.) S+SSPR 2018. LNCS, vol. 11004, pp. 326–336. Springer, Cham (2018). https://doi.org/10.1007/978-3-319-97785-0_31
8. Fischer, A., Riesen, K., Bunke, H.: Improved quadratic time approximation of graph edit distance by combining hausdorff matching and greedy assignment. Pattern Recognit. Lett. **87**, 55–62 (2017)
9. Riesen, K., Bunke, H.: Approximate graph edit distance computation by means of bipartite graph matching. Image Vis. Comput. **27**(7), 950–959 (2009)
10. Serratosa, F., Cortés, X.: Interactive graph-matching using active query strategies. Pattern Recognit. **48**(4), 1364–1373 (2015)
11. Ferrer, M., Serratosa, F., Riesen, K.: Improving bipartite graph matching by assessing the assignment confidence. Pattern Recognit. Lett. **65**, 29–36 (2015)
12. Serratosa, F.: Speeding up fast bipartite graph matching through a new cost matrix. Int. J. Pattern Recognit. Artif. Intell. **29**(02), 1550010 (2015)
13. Munkres, J.: Algorithms for the assignment and transportation problems. J. Soc. Ind. Appl. Math. **5**(1), 32–38 (1957)
14. Cortés, X., Serratosa, F., Riesen, K.: On the relevance of local neighbourhoods for greedy graph edit distance. In: Robles-Kelly, A., Loog, M., Biggio, B., Escolano, F., Wilson, R. (eds.) S+SSPR 2016. LNCS, vol. 10029, pp. 121–131. Springer, Cham (2016). https://doi.org/10.1007/978-3-319-49055-7_11

15. Serratosa, F., Cortés, X.: Graph edit distance: moving from global to local structure to solve the graph-matching problem. Pattern Recognit. Lett. **65**, 204–210 (2015)

16. Fischer, A., Riesen, K., Horst, B.: Improved quadratic time approximation of graph edit distance by combining Hausdorff matching and greedy assignment. Pattern Recognit. Lett. **87**, 55–62 (2017)

17. Riesen, K., Ferrer, M., Dornberger, R., Bunke, H.: Greedy graph edit distance. In: Perner, P. (ed.) MLDM 2015. LNCS (LNAI), vol. 9166, pp. 3–16. Springer, Cham (2015). https://doi.org/10.1007/978-3-319-21024-7_1

18. Serratosa, F.: Fast computation of bipartite graph matching. Pattern Recognit. Lett. **45**, 244–250 (2014)

19. Jonker, R., Volgenant, T.: Improving the Hungarian assignment algorithm. Oper. Res. Lett. **5**(4), 171–175 (1986)

20. Caetano, T.S., McAuley, J.J., Cheng, L., Le, Q.V., Smola, A.J.: Learning graph matching. IEEE Trans. PAMI **31**(6), 1048–1058 (2009)

21. Leordeanu, M., Sukthankar, R., Hebert, M.: Unsupervised learning for graph matching. Int. J. Comput. Vis. **96**(1), 28–45 (2012)

22. Cortés, X., Serratosa, F.: Learning graph-matching edit-costs based on the optimality of the Oracle's node correspondences. Pattern Recognit. Lett. **56**, 22–29 (2015)

23. Cortés, X., Serratosa, F.: Learning graph matching substitution weights based on the ground truth node correspondence. IJPRAI **30**(02), 1650005 (2016)

24. Algabli, S., Serratosa, F.: Embedding the node-to-node mappings to learn the graph edit distance parameters. Pattern Recognit. Lett. **112**, 353–360 (2018)

25. Neuhaus, M., Bunke, H.: Automatic learning of cost functions for graph edit distance. Inf. Sci. **177**(1), 239–247 (2007)

26. Neuhaus, M., Bunke, H.: Self-organizing maps for learning the edit costs in graph matching. IEEE Trans. SMC, Part B **35**(3), 503–514 (2005)

27. Dunn, J.C.: Well-separated clusters and optimal fuzzy partitions. J. Cybern. **4**(1), 95–104 (1974)

28. Moreno-García, C.F., Cortés, X., Serratosa, F.: A graph repository for learning error-tolerant graph matching. In: Proceedings of SSPR (2016)

29. Serratosa, F.: A methodology to generate attributed graphs with a bounded graph edit distance for graph-matching testing. IJPRAI **32**(11), 1850038 (2018)

Cross-Evaluation of Graph-Based Keyword Spotting in Handwritten Historical Documents

Michael Stauffer[1][(✉)] [ID], Paul Maergner[2] [ID], Andreas Fischer[2,3] [ID],
and Kaspar Riesen[1] [ID]

[1] Institute for Information Systems,
University of Applied Sciences and Arts Northwestern Switzerland,
Riggenbachstrasse 16, 4600 Olten, Switzerland
{michael.stauffer,kaspar.riesen}@fhnw.ch
[2] Department of Informatics, University of Fribourg,
Boulevard de Pérolles 90, 1700 Fribourg, Switzerland
{paul.maergner,andreas.fischer}@unifr.ch
[3] Institute of Complex Systems,
University of Applied Sciences and Arts Western Switzerland,
Boulevard de Pérolles 80, 1700 Fribourg, Switzerland

Abstract. In contrast to statistical representations, graphs offer some inherent advantages when it comes to handwriting representation. That is, graphs are able to adapt their size and structure to the individual handwriting and represent binary relationships that might exist within the handwriting. We observe an increasing number of graph-based keyword spotting frameworks in the last years. In general, keyword spotting allows to retrieve instances of an arbitrary query in documents. It is common practice to optimise keyword spotting frameworks for each document individually, and thus, the overall generalisability remains somehow questionable. In this paper, we focus on this question by conducting a cross-evaluation experiment on four handwritten historical documents. We observe a direct relationship between parameter settings and the actual handwriting. We also propose different ensemble strategies that allow to keep up with individually optimised systems without *a priori* knowledge of a certain manuscript. Such a system can potentially be applied to new documents without prior optimisation.

Keywords: Keyword spotting · Handwritten historical documents ·
Graph-based representations · Hausdorff Edit Distance ·
Ensemble methods

1 Introduction

Different handwritten historical documents often show large variations in the handwriting (e.g. scale or style) and are often negatively affected by ink-bleed through, fading, etc. Consequently, an automatic full transcription is often not

© Springer Nature Switzerland AG 2019
D. Conte et al. (Eds.): GbRPR 2019, LNCS 11510, pp. 45–55, 2019.
https://doi.org/10.1007/978-3-030-20081-7_5

feasible [3]. For this reason, *Keyword Spotting (KWS)* has been proposed as a more flexible and error-tolerant alternative [5]. In particular, KWS systems allow to retrieve all word instances in handwritten historical documents that represent a given query word.

1.1   Related Work

In graph-based KWS, a query graph is commonly matched with the graphs that represent the document words. Hence, sorted graph dissimilarities can be used to derive a retrieval index that consists – in the best case – of all relevant keywords as its top results.

Different graph-based approaches for KWS are based on different representations of the handwriting. However, nodes are often used to represent characteristic points (so called *keypoints*) in the handwriting, while edges are commonly used to represent handwriting strokes [13]. In other approaches the nodes are used to represent prototype strokes, while edges are used to connect nodes that stem from the same connected component [2,8]. More recently, a set of different graph-based handwriting representations has been proposed that make use of keypoints, grid-wise segmentations, or projection profiles [10]. These handwriting graph representations have been actually employed in various graph-based KWS applications [1,7,11,12]. Very recently, Deep Learning techniques (so called *Message Passing Neural Networks*) have been used to enhance node labels by a structural node context [7].

Regardless the graph representation actually used, a matching procedure is required in order to conduct KWS. To this end, different graph dissimilarities have been employed like, for instance, *Bipartite Graph Edit Distance (BP)* [2,8,11–13][1] as well as *Hausdorff Edit Distance (HED)* [1,7][2]. Moreover, *ensemble methods* have been proposed to combine different graph representations [11].

1.2   Contribution

It is common practice in the field of KWS research that parameters are individually optimised for every document [2,3,5,7,8,13]. That is, the parameters are often optimised on a subset of a specific document and then tested on a disjoint set stemming from the same document. However, this practice does not reflect a realistic scenario especially as libraries often keep thousands of different handwritten historical documents. It would be a very cumbersome and time consuming task to individually optimise a given KWS system for each of these documents.

In the present paper, we evaluate the generalisability of a graph-based KWS system. That is, we investigate the performance and limitation of this system in a cross-evaluation experiment on four handwritten historical documents,

---

[1] BP has been introduced in [9].
[2] HED has been introduced in [4].

(a)   (b)

(c)   (d)

Fig. 1. Exemplary excerpts of four handwritten historical documents: (a) George Washington (GW), (b) Parzival (PAR), (c) Alvermann Konzilsprotokolle (AK), (d) Botany (BOT).

viz. *George Washington (GW)*[3], *Parzival (PAR)*[4], *Alvermann Konzilsprotokolle (AK)*, and *Botany (BOT)*[5]. In particular, we optimise parameters on one document (for instance GW) and eventually test the optimised settings on the three remaining documents (in this case PAR, AK, and BOT). We repeat this procedure for each document. Moreover, we propose and evaluate novel ensemble methods that allow to test unknown documents without prior optimisation step. That is, these ensemble systems combine the results of three KWS systems (individually optimised on three different manuscripts) in order to instantly perform KWS on an unseen document.

In Fig. 1, excerpts from each document are shown. The large variations in the writing styles and document states are clearly visible and illustrate the challenging task of tuning a KWS system on one document that eventually returns reasonable results on other documents.

The remainder of this paper is organised as follows. First, the graph-based KWS framework actually employed for our research study is reviewed in Sect. 2. Next, the cross-evaluation experiment on the four handwritten documents as well as the ensemble results are presented and discussed in Sect. 3. Finally, we draw conclusions and discuss further research activities in Sect. 4.

2   Graph-Based Keyword Spotting

In this section, we review a graph-based KWS framework originally proposed in [1,12]. We use this framework as basic system to conduct both the cross validation and the ensemble experiments. This framework is based on three different

---

[3] George Washington Papers at the Library of Congress, 1741–1799: Series 2, Letterbook 1, pp. 270–279 & 300–309, http://memory.loc.gov/ammem/gwhtml/gwseries2.html.

[4] Parzival at IAM historical document database, http://www.fki.inf.unibe.ch/databases/iam-historical-document-database/parzival-database.

[5] Alvermann Konzilsprotokolle and Botany at ICFHR2016 benchmark database, http://www.prhlt.upv.es/contests/icfhr2016-kws/data.html.

processing steps (as illustrated in Fig. 2) and is briefly outlined in the next three subsections. In the fourth and last subsection we discuss a possibility to build an ensemble out of different KWS systems that might be particularly useful in order to increase the generalisability of a KWS system.

2.1   Image Preprocessing

For the two documents GW and PAR, general noise is addressed by means of *Difference of Gaussians* filtering. Next, document images are binarised by global thresholding. Moreover, the resulting document images are automatically segmented into single word images by means of their projection profiles, and if necessary manually corrected. That is, we focus on the KWS process itself and assume perfectly segmented documents in our evaluations. For deskewing, the angle between x-axis and lower baseline of a text line is estimated and used to rotate single word images. Finally, preprocessed word images are skeletonised by means of thinning.

For the two documents AK and BOT, segmented word images are directly taken from the ICFHR2016 benchmark database [6], and thus, only binarisation has been employed. To handle small segmentation errors, we employ an additional image preprocessing step that removes small connected components on these two manuscripts.

We denote preprocessed and skeletonised word images by S from now on. For more details on the preprocessing step we refer to [11,12].

2.2   Handwriting Graphs

In general, a graph g is defined as a four-tuple $g = (V, E, \mu, \nu)$ where V and E are finite sets of nodes and edges, and $\mu : V \to L_V$ and $\nu : E \to L_E$ are labelling functions for nodes and edges, respectively. The handwriting graphs employed in this paper are defined as follows. Nodes are used to represent characteristic points, so-called *keypoints*, in the handwriting, while edges are used to represent strokes between keypoints. Hence, nodes are labelled with two-dimensional numerical labels, while edges remain unlabelled, i.e. $L_V = \mathbb{R}^2$ and $L_E = \emptyset$. In the following paragraphs we briefly review the procedure of extracting graphs from word images (for details we refer to [12]).

First, end points and junction points are identified in the word images S. Selected keypoints are added to the graph as nodes and labelled with their respective (x, y)-coordinates. Next, intermediate points are added as nodes along the skeleton in equidistant intervals of size D. Eventually, an undirected edge (u, v) between $u \in V$ and $v \in V$ is inserted into the graph for each pair of nodes that is directly connected by a chain of foreground pixels in image S.

To reduce scaling variations, the (x, y)-coordinates of the node labels $\mu(v)$ are normalised by a z-score. Formally, we replace (x, y) by $(\hat{x}, \hat{y})$, where

$$\hat{x} = \frac{x - \mu_x}{\sigma_x} \text{ and } \hat{y} = \frac{y - \mu_y}{\sigma_y}.$$

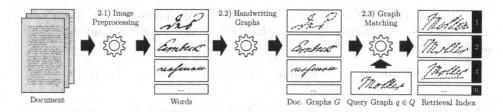

Fig. 2. Graph-based keyword spotting processing of the word "Möller".

Thereby (μ_x, μ_y) and (σ_x, σ_y) represent the mean and standard deviation of all (x, y)-coordinates in the graph under consideration[6].

For each manuscript, an original word image, a preprocessed word image, a skeletonised word image, as well as the corresponding handwriting graph is given in Fig. 3.

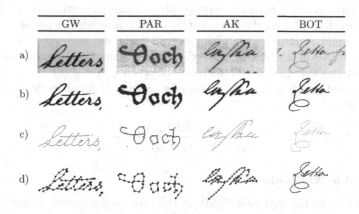

Fig. 3. Exemplary graph representation of four handwritten historical documents (viz. George Washington (GW), Parzival (PAR), Alvermann Konzilsprotokolle (AK), and Botany (BOT)): (a) Original word image, (b) Preprocessed word image, (d) Skeletonised word image, (c) Handwriting graph.

2.3 Graph Matching

The actual keyword spotting is based on a pairwise matching of a query graph q with all graphs g stemming from the set of document graphs G. In this paper, we make use of *Hausdorff Edit Distance (HED)* [4]. HED is a quadratic time lower bound of *Graph Edit Distance* that measures the minimum-cost deformation needed to transform one graph $g_1 = (V_1, E_1, \mu_1, \nu_1)$ into another graph $g_2 = (V_2, E_2, \mu_2, \nu_2)$ by means of deletions $(u \rightarrow \epsilon)$, insertions $(\epsilon \rightarrow v)$, and

---

[6] Note that the resulting graphs are available under http://www.histograph.ch/.

substitutions $(u \to v)$ of nodes $u \in V_1$ and $v \in V_2$. Likewise, edit operations are defined for the edges. Formally, the HED of two graphs g_1 and g_2 can be derived by

$$\text{HED}(g_1, g_2) = \sum_{u \in V_1} \min_{v \in V_2 \cup \{\epsilon\}} f(u, v) + \sum_{v \in V_2} \min_{u \in V_1 \cup \{\epsilon\}} f(u, v),$$

where $f(u, v)$ is a cost function that takes into account the node edit cost $c(u \to v)$ as well as the edge edit cost $c(q \to r)$ for all edges q and r adjacent to u and v, respectively.

The cost model employed is based on a constant cost $\tau_v \in \mathbb{R}^+$ for node deletions/insertions and a constant cost $\tau_e \in \mathbb{R}^+$ for edge deletions/insertions. For node substitutions, the following weighted Euclidean distance is employed:

$$\sqrt{\alpha \left(\sigma_x (x_i - x_j)\right)^2 + (1 - \alpha) \left(\sigma_y (y_i - y_j)\right)^2},$$

where $\alpha \in [0, 1]$ denotes a parameter to weight the importance of the x- and y-coordinate of a node, while σ_x and σ_y denote the standard deviation of all node coordinates in the current query graph q. Edge substitutions are free of cost (since they are unlabelled). We additionally use a weighting factor $\beta \in [0, 1]$ to weight the relative importance of the overall node and edge edit costs.

Finally, a retrieval index r is derived. In particular, HED is normalised by the maximum possible graph edit distance between q and g (i.e. the sum that results from deleting all nodes and edges of q and inserting all nodes and edges in g). Formally,

$$r(q, g) = \frac{\text{HED}(q, g)}{(|V_q| + |V_g|) \, \tau_v + (|E_q| + |E_g|) \, \tau_e}.$$

2.4 Ensemble Methods

In order to increase the generalisability of the proposed framework, we propose three different ensemble methods that allow to combine optimised cost models of known documents. The general idea of these systems is as follows. We assume that we have three documents at hand on which a KWS system can be individually optimised. We eventually apply all three parametrisations to one unknown document and combine the three results by means of a statistical measure. Formally,

$$r_{\min}(q, g) = \min_{i \in \{A, B, C\}} r_i(q, g),$$

$$r_{\max}(q, g) = \max_{i \in \{A, B, C\}} r_i(q, g),$$

$$r_{\text{mean}}(q, g) = \underset{i \in \{A, B, C\}}{\text{mean}} \, r_i(q, g),$$

where $\{A, B, C\}$ represent three given manuscripts, and r_i refers to the HED optimised on manuscript A, B, or C. If we assume, for instance, that BOT is an unknown document, $\{A, B, C\}$ is given by the three remaining documents that is $A = \text{GW}$, $B = \text{PAR}$, and $C = \text{AK}$.

3   Experimental Evaluation

3.1   Experimental Setup

For all evaluations, the accuracy is measured by the *Mean Average Precision (MAP)*, which is the mean area under all recall-precision curves of all individual keywords. In particular, the evaluation is conducted in two steps, viz. validation and test.

First, ten different keywords (with different word lengths) are manually selected on each dataset. Based on these keywords, we define an independent validation set for parameter optimisation that consists of 10 random instances per keyword instance and 900 additional random words (in total 1,000 words). We evaluate 25 pairs of constants for node and edge deletion/insertion costs ($\tau_v = \tau_e \in \{1, 4, 8, 16, 32\}$) in combination with the weighting parameters $\alpha = \beta \in \{0.1, 0.3, 0.5, 0.7, 0.9\}$ (see Sect. 2.3). In Table 1, the optimal cost model for each manuscript is given.

Table 1. Optimal cost function parameter.

GW				PAR				AK				BOT			
τ_v	τ_e	α	β	τ_v	τ_e	α	β	τ_v	τ_e	α	β	τ_v	τ_e	α	β
8	4	0.1	0.5	8	1	0.5	0.1	16	1	0.1	0.3	8	4	0.1	0.3

Next, the proposed framework is tested using the same training and test sets as used in [3] and [6]. In Table 2, a summary of dataset characteristics of all four documents is given.

Table 2. Number of keywords, size of keyword spotting datasets (train and test), and the image resolution of the original documents in dpi.

Dataset	Keywords	Train	Test	dpi
GW	105	2447	1224	300
PAR	1217	11468	6869	200
AK	200	1849	3734	400
BOT	150	1684	3380	400

3.2   Cross-Evaluation

In Table 3, the results of the cross-evaluation are shown for all manuscripts (columns) using cross-evaluated parameters (rows). For instance, in the first row we show the KWS results on all four data sets of the system actually optimised on GW. In the main diagonal of Table 3 we thus provide the KWS results on a document obtained by the system optimised on the same document.

On GW we observe that the KWS system optimised on BOT achieves a similar result as the document specific system (actually this system performs even slightly better). Regarding the optimal parameters for the cost function in Table 1, this result makes sense as for both GW and BOT very similar parameters turn out to be optimal. Also the writing styles of both documents are quite similar (see, for instance, Fig. 1).

Likewise, we observe that on BOT the KWS system optimised on GW achieves quite similar results as the system optimised on BOT itself. In contrast with GW, however, we observe that on this dataset the parametrisation seems to have less influence on the KWS accuracy as all parametrisations lead to similar results. The same accounts for AK, where the optimal parameters for BOT turn out to achieve the best result on the test set.

On one document, viz. PAR, however, none of the systems optimised on another document can actually keep up with the system that has been optimised for this specific manuscript. That is we observe deteriorations of the KWS accuracy of about 6 to 12 basis points. The system optimised on PAR makes use of $\alpha = 0.5$, while all other systems turn out to be optimal with $\alpha = 0.1$. PAR has a more dense and straight (i.e. almost no slant) handwriting when compared to GW, AK, and BOT as shown in Fig. 1. As a result, variations in the x-direction become more relevant (thus the higher α value).

Table 3. MAP using optimised cost function parameters of one manuscript employed on all three remaining manuscripts. With ± we indicate the absolute percental gain or loss in the accuracy of the cross-evaluated manuscript when compared with the optimised parameter settings (shown in bold face).

Optimised on	GW		PAR		AK		BOT	
	MAP	±	MAP	±	MAP	±	MAP	±
GW	**69.28**	-	63.39	−5.84	79.54	−0.18	51.21	−0.48
PAR	64.84	−4.44	**69.23**	-	79.73	+0.01	51.12	−0.57
AK	61.45	−7.82	56.93	−12.30	**79.72**	-	50.81	−0.88
BOT	69.44	+0.17	62.40	−6.83	80.28	+0.56	**51.69**	-

Overall we conclude, that the weighting parameter α shows quite a strong correlation with the density of the handwriting. For a dense and straight handwriting the x-direction becomes more important, and thus higher parameter values for α should be chosen. In contrast to that, β has in most cases only a minor influence. Finally, it seems that the cost parameters τ_v and τ_e are depending on the size of the handwriting. That is, if the handwriting is characterised by flourish like in case of AK, for instance, node substitutions should be rather allowed by the cost model (by defining higher values for τ_v).

3.3 Ensemble Methods

In Table 4, we show the results of the proposed ensemble methods and the individually optimised systems for each document. In the first column, for instance, we show in the first row the KWS accuracy on GW of the system actually optimised on GW. The three ensemble methods combine the results of the three systems optimised on the remaining datasets.

In three out of four manuscripts, we observe that the ensemble methods can keep up or even improve the accuracy when compared with the individually optimised system. Especially, the ensemble methods max and mean achieve similar KWS accuracies without any a priori knowledge of the manuscript. Similar to the cross-evaluation experiment, we observe that ensemble methods can not keep up on PAR. In contrast to all other manuscripts PAR offers a different writing style, and the ensemble methods are not able to compensate these obvious differences.

Table 4. MAP of ensemble methods min, max, and mean. With $\pm$ we indicate the absolute percental gain or loss in the accuracy of the ensemble method when compared with the optimised parameter settings. Ensemble methods are ranked by (1), (2), (3).

	GW		PAR		AK		BOT		Average	
	MAP	$\pm$	MAP	$\pm$	MAP	$\pm$	MAP	$\perp$	MAP	$\perp$
Optimal	69.28		69.23		79.72		51.69		67.48	
min	65.63	−3.65 (3)	56.94	−12.29 (3)	80.28	+0.56 (2)	50.81	−0.88 (3)	63.42	−4.06 (3)
max	69.25	−0.02 (1)	63.39	−5.84 (1)	79.54	−0.18 (3)	51.25	−0.44 (2)	65.86	−1.62 (1)
mean	66.65	−2.62 (2)	62.28	−6.95 (2)	80.42	+0.70 (1)	51.49	−0.20 (1)	65.21	−2.27 (2)

We conclude that with ensemble methods (without document specific adaptations) similar KWS accuracy rates can be achieved as with individual document specific systems in most cases.

4 Conclusion and Outlook

The automatic recognition of handwritten historical documents is often negatively affected by noise (ink-bleed through, fading, degradation, etc.) as well as large handwriting variations in different documents. Therefore, *Keyword Spotting (KWS)* has been proposed as flexible and error-tolerant alternative to full transcriptions. Basically keyword spotting allows arbitrary retrievals in a document in order to make such a document accessible for browsing and searching.

In the last years, a number of graph-based keyword spotting approaches have been proposed. Yet, all of the proposed approaches are individually optimised and tested for each manuscript. Consequently, for a novel unseen document the system first needs to be optimised prior to the actual keyword spotting. In case of large collections or libraries this clearly reduces the overall applicability and practical relevance of the proposed graph-based keyword spotting frameworks.

In order to research this problem we conduct a cross-evaluation on four handwritten historical documents. That is, we evaluate KWS systems that have been optimised on other, unrelated documents. We observe a clear relationship between handwriting style and cost model for graph edit distance. Therefore, an unseen document could be directly accessed by a KWS system that has been optimised on a similar document without *a priori* parameter optimisation. Moreover, we show that ensemble methods allow to further increase the overall generalisability of the graph-based KWS. That is, the proposed ensemble methods, that need no document specific training, achieve similar accuracy rates as the optimised cost models.

In future work we aim at including further documents with different handwriting styles to our evaluation pipeline. Another research avenue to be pursued would be to research an automatic a priori *triage* in order to sort unknown manuscripts by means of their handwriting style. Finally, one could also extend the proposed ensemble methods by means of so-called *overproduce-and-select strategies*. That is, starting with the best individual system further cost model settings are added to an ensemble until a certain saturation is reached.

Acknowledgements. This work has been supported by the Swiss National Science Foundation project 200021_162852.

References

1. Ameri, M.R., Stauffer, M., Riesen, K., Bui, T.D., Fischer, A.: Graph-based keyword spotting in historical manuscripts using Hausdorff edit distance. Pattern Recognit. Lett. (2018, in press)
2. Bui, Q.A., Visani, M., Mullot, R.: Unsupervised word spotting using a graph representation based on invariants. In: International Conference on Document Analysis and Recognition, pp. 616–620. IEEE (2015)
3. Fischer, A., Keller, A., Frinken, V., Bunke, H.: Lexicon-free handwritten word spotting using character HMMs. Pattern Recognit. Lett. **33**(7), 934–942 (2012)
4. Fischer, A., Suen, C.Y., Frinken, V., Riesen, K., Bunke, H.: Approximation of graph edit distance based on Hausdorff matching. Pattern Recognit. **48**(2), 331–343 (2015)
5. Manmatha, R., Han, C., Riseman, E.: Word spotting: a new approach to indexing handwriting. In: Computer Society Conference on Computer Vision and Pattern Recognition, pp. 631–637. IEEE (1996)
6. Pratikakis, I., Zagoris, K., Gatos, B., Puigcerver, J., Toselli, A.H., Vidal, E.: ICFHR2016 handwritten keyword spotting competition (H-KWS 2016). In: International Conference on Frontiers in Handwriting Recognition, pp. 613–618. IEEE (2016)
7. Riba, P., Fischer, A., Lladós, J., Fornés, A.: Learning graph distances with message passing neural networks. In: International Conference on Pattern Recognition. IEEE (2018)
8. Riba, P., Lladós, J., Fornés, A.: Handwritten word spotting by inexact matching of grapheme graphs. In: International Conference on Document Analysis and Recognition, pp. 781–785. IEEE (2015)

9. Riesen, K., Bunke, H.: Approximate graph edit distance computation by means of bipartite graph matching. Image Vis. Comput. **27**(7), 950–959 (2009)
10. Stauffer, M., Fischer, A., Riesen, K.: A novel graph database for handwritten word images. In: Robles-Kelly, A., Loog, M., Biggio, B., Escolano, F., Wilson, R. (eds.) S+SSPR 2016. LNCS, vol. 10029, pp. 553–563. Springer, Cham (2016). https://doi.org/10.1007/978-3-319-49055-7_49
11. Stauffer, M., Fischer, A., Riesen, K.: Ensembles for graph-based keyword spotting in historical handwritten documents. In: International Conference on Document Analysis and Recognition, pp. 714–720. IEEE (2017)
12. Stauffer, M., Fischer, A., Riesen, K.: Keyword spotting in historical handwritten documents based on graph matching. Pattern Recognit. **81**, 240–253 (2018)
13. Wang, P., Eglin, V., Garcia, C., Largeron, C., Lladós, J., Fornés, A.: A novel learning-free word spotting approach based on graph representation. In: International Workshop on Document Analysis Systems, pp. 207–211. IEEE (2014)

Graph Edge Entropy
in Maxwell-Boltzmann Statistics
for Alzheimer's Disease Analysis

Jianjia Wang[1]([✉]), Richard C. Wilson[2], and Edwin R. Hancock[2]

[1] School of Computer Science, Shanghai University, Shanghai, China
jianjiawang@shu.edu.cn
[2] Department of Computer Science, University of York, York, UK

Abstract. In this paper, we explore how to the decompose the global thermodynamic entropy of a network into components associated with its edges. Commencing from a statistical mechanical picture in which the normalised Laplacian matrix plays the role of Hamiltonian operator, thermodynamic entropy can be calculated from partition function associated with different energy level occupation distributions arising from Maxwell-Boltzmann statistics. Using the spectral decomposition of the Laplacian, we show how to project the edge-entropy components so that the detailed distribution of entropy across the edges of a network can be achieved. We apply the resulting method to fMRI activation networks to evaluate the qualitative and quantitative characterisations. The entropic measurement turns out to be an effective tool to identify the differences in the structure of Alzheimer's disease by selecting the most salient anatomical brain regions.

Keywords: fMRI networks · Graph edge entropy ·
Alzheimer's disease (AD)

1 Introduction

Functional magnetic resonance imaging (fMRI) provides a sophisticated means of studying the neuropathophysiology associated with Alzheimer's disease (AD) [1]. It maps the network representation to neuronal activity between the various brain regions. The resulting network structure has proved useful in understanding Alzheimer's disease (AD) via the analysis of intrinsic brain connectivity [2]. Although there are many tools to identify the affected brain regions in AD, it is still not clear how this abnormality affects the functional organization of the whole brain connection.

Tools from network analysis provide a convenient approach for understanding the functional association of different regions in the brain [2,4]. The approach is to characterize the topological structures present in the brain and to quantify the functional interaction between brain regions, using the mathematical study of networks and graph theory. Graph theory offers an attractive route since it provides effective tools for characterizing network structures together with

© Springer Nature Switzerland AG 2019
D. Conte et al. (Eds.): GbRPR 2019, LNCS 11510, pp. 56–66, 2019.
https://doi.org/10.1007/978-3-030-20081-7_6

their intrinsic complexity [3,6]. This approach has led to the design of several practical methods for characterizing the global and local structure of graphs [3]. Features based on the global and local measures of connectivity are widely used in functional brain analysis [5].

Unfortunately, there is relatively little literature aimed at studying structural network features using entropic analysis. The reason for that is the vast majority of techniques suggested by entropy provides a useful global characterisation of network structure, they do not lend themselves to the analysis of local measures of edge or subnetwork structure. However, entropy analysis is a more natural representation for brain structure, since they allow the information of activation signals for different anatomical structures in the brain.

This paper is motivated by the need to fill this important gap in the literature, and to establish effective methods for measuring the structural properties of entropy representing fMRI brain networks. In particular, in order to characterize the functional organization of the brain, our approach explores a novel edge entropy projection which can decompose the global network entropy computed from Maxwell-Boltzmann distribution [6,12]. The new characterisations of edge entropy resulting from this analysis allow us to probe in finer detail the interactions between different anatomical regions in fMRI data from healthy controls and Alzheimer's disease sufferers (AD).

The remainder of the paper is organized as follows. Section 2 briefly reviews the basic concepts in network representation, especially with a sophisticated study of von Neumann entropy. Section 3 reviews density matrix and Hamiltonian operator on graphs, and decompose the thermodynamic entropy on edges from Maxwell-Boltzmann statistics. Section 4 provides our experimental evaluation. Finally, Sect. 5 provides the conclusion and direction for future work.

2 Graph Representation

2.1 Preliminaries

Let $G(V, E)$ be an undirected graph with node set V and edge set $E \subseteq V \times V$, and let V represent the total number of nodes on graph $G(V, E)$. The adjacency matrix of a graph is A with the degree of node u is $d_u = \sum_{v \in V} A_{uv}$. Then, the Laplacian matrix is $L = D - A$, where D denotes the degree diagonal matrix whose elements are given by $D(u, u) = d_u$ and zeros elsewhere. The normalized Laplacian matrix $\tilde{L}$ of the graph G is defined as $\tilde{L} = D^{-\frac{1}{2}} L D^{\frac{1}{2}}$, and the spectral decomposition is $\tilde{L} = \Phi \tilde{\Lambda} \Phi^T$, where $\tilde{\Lambda} = diag(\lambda_1, \lambda_2, \ldots \lambda_{|V|})$ is the diagonal matrix with the ordered eigenvalues as elements and $\Phi = (\varphi_1, \varphi_2, \ldots, \varphi_{|V|})$ is the matrix with the ordered eigenvectors as columns.

2.2 Von Neumann Edge Entropy

In quantum mechanics, the density matrix is used to describe a system with the probability of pure quantum states $|\psi_i\rangle$ and each with probability p_i. It is defined as $\rho = \sum_{i=1}^{V} p_i |\psi_i\rangle\langle\psi_i|$. Severini et al. [7] have extended this idea to the graph domain. Specifically, they show that a density matrix for a graph or network can be obtained by scaling the combinatorial Laplacian matrix by the reciprocal of the number of nodes in the graph.

With this notation, the specified density matrix is obtained by scaling the normalized Laplacian matrix by the number of nodes, i.e. $\rho = \frac{\tilde{L}}{|V|}$. When defined in this way the density matrix is Hermitian i.e. $\rho = \rho\dagger$ and $\rho \geq 0, \mathrm{Tr}\rho = 1$. This interpretation opens up the possibility of characterising a graph using the von Neumann entropy from quantum information theory [7]. Therefore, the von Neumann entropy is given in terms of the eigenvalues $\lambda_1, \ldots, \lambda_{|V|}$ of the density matrix ρ,

$$S_{VN} = -\mathrm{Tr}(\rho \log \rho) = -\sum_{i=1}^{|V|} \frac{\lambda_i}{|V|} \log \frac{\lambda_i}{|V|} \tag{1}$$

In fact, Han et al. [8] have shown how to approximate the calculation of von Neumann entropy in terms of simple degree statistics. Their approximation allows the cubic complexity of computing the von Neumann entropy to be reduced to one of quadratic complexity using simple edge degree statistics, i.e.

$$S_{VN} = 1 - \frac{1}{|V|} - \frac{1}{|V|^2} \sum_{(u,v)\in E} \frac{1}{d_u d_v} \tag{2}$$

Therefore, the edge entropy decomposition is given as

$$S_{VN}^{edge}(u,v) = \frac{1}{|E|} - \frac{1}{|V||E|} - \frac{1}{|E||V|^2} \frac{1}{d_u d_v} \tag{3}$$

where $S_{VN} = \sum_{(u,v)\in E} S_{VN}^{edge}(u,v)$. This expression decomposes the global parameter of von Neumann entropy on each edge with the relation to the degrees from the connection of two vertexes.

3   Thermodynamic Statistics and Global Entropy Decomposition

The concept of von Neumann entropy arises in the quantum domain. Here, we commence from the Hamiltonian operator in statistical mechanics to develop thermodynamic entropy. We then decompose or project the global entropy onto edges using the eigenvectors of the normalised Laplacian matrix.

3.1   Thermodynamic Entropy

To connect the normalised Laplacian matrix to statistical mechanics, we view the eigenvalues of the Laplacian matrix as the energy eigenstates of a system in contact with a heat reservoir. These determine the Hamiltonian and hence the relevant Schrödinger equation which governs the particles in the system [3,6].

The particles occupy the energy states of the Hamiltonian subject to thermal agitation by the heat bath [12]. The number of particles in each energy state is determined by the temperature, the assumed model of occupation statistics and the relevant chemical potential.

We consider the network as a thermodynamic system of N particles with energy states given by normalised Laplacian matrix $\tilde{L}$, which is immersed in a heat bath with temperature T. The ensemble is represented by a partition function $Z(\beta, N)$, where β is inverse of temperature T. When specified in this way, the thermodynamic entropy is given by,

$$S = k_B \left[\frac{\partial}{\partial T} T \log Z \right]_N \tag{4}$$

The statistical properties of particles in the network are determined by the partition functions associated with different energy level occupation statistics. In this way, thermodynamic quantities, such as entropy, can characterise the network structure.

3.2   Maxwell-Boltzmann Statistics

The Maxwell-Boltzmann distribution relates the microscopic properties of particles to the macroscopic thermodynamic properties of matter [10]. It applies to systems consisting of a fixed number of weakly interacting distinguishable particles. These particles occupy the energy levels associated with a Hamiltonian and in our case the Hamiltonian of the network, which is in contact with a thermal bath [6].

Taking the Hamiltonian to be the normalized Laplacian of the network, the canonical partition function for Maxwell-Boltzmann occupation statistics of the energy levels is

$$Z_{MB} = \mathrm{Tr}\left[\exp(-\beta \tilde{L})^N \right] \tag{5}$$

where $\beta = 1/k_B T$ is the reciprocal of the temperature T with k_B as the Boltzmann constant; N is the total number of particles and λ_i denotes the microscopic energy of system at each microstate i with energy λ_i. Derived from Eq. (4), the entropy of the system with N particles is

$$S_{MB} = \log Z - \beta \frac{\partial \log Z}{\partial \beta} = -N \mathrm{Tr}\left\{ \frac{\exp(-\beta \tilde{L})}{\mathrm{Tr}[\exp(-\beta \tilde{L})]} \log \frac{\exp(-\beta \tilde{L})}{\mathrm{Tr}[\exp(-\beta \tilde{L})]} \right\}$$

For a single particle, the density matrix is

$$\rho_{MB} = \frac{\exp(-\beta \tilde{L})}{\mathrm{Tr}[\exp(-\beta \tilde{L})]} \tag{6}$$

Since the density matrix commutes with the Hamiltonian operator, we have $\partial \rho / \partial t = 0$ and the system can be viewed as in equilibrium. So the entropy in the Maxwell-Boltzmann system is simply N times the von Neumann entropy of a single particle, as we might expect.

3.3 Edge Entropy Decomposition

Our goal is to project the global network entropy onto the edges of the network. In matrix form for Maxwell-Boltzmann statistics in Eq. (6), the entropy can be written as,

$$S_{MB} = -\text{Tr}\left[\rho_{MB} \log \rho_{MB}\right] = -\text{Tr}[\Sigma_{MB}] \tag{7}$$

Since the spectral decomposition of the normalized Laplacian matrix is

$$\tilde{L} = \Phi\tilde{\Lambda}\Phi^T \tag{8}$$

We can decompose the matrix Σ_{MB} as follows

$$\Sigma_{MB} = \Phi\sigma_{MB}(\tilde{\Lambda})\Phi^T \tag{9}$$

where

$$\sigma_{MB}(\lambda_i) = -N\frac{e^{-\beta\lambda_i}}{\sum_{i=1}^{|V|} e^{-\beta\lambda_i}} \log \frac{e^{-\beta\lambda_i}}{\sum_{i=1}^{|V|} e^{-\beta\lambda_i}} \tag{10}$$

As a result, we can perform edge entropy projection of the Maxwell-Boltzmann statistical model using the Laplacian eigenvectors [11]. The result of the entropy for each edge (uv) is given as,

$$S_{MB}^{edge}(u,v) = \sum_{i=1}^{|V|} \sigma_{MB}(\lambda_i)\varphi_i\varphi_i^T \tag{11}$$

Thus, the global entropy can be projected on the edges of the network system. This provides useful measures for local entropic characterisation of network structure in a relatively straightforward manner.

4　Experiments and Evaluations

In this section, we describe the application of the above methods to the analysis of interregional connectivity structure for fMRI activation networks for normal and Alzheimer's patients. We first examine the dependence of the edge entropy components on node degree and temperature and compare their performance with von Neumann entropy. Then we apply edge entropy-based analysis to distinguish between different stages in the development of Alzheimer's disease, and fMRI data for normal subjects. We explore whether we can identify specific interregional connections and regions in the brain associated with the neurodegeneration caused by the onset of Alzheimer's disease. To simplify the calculations, the Boltzmann constant is set to unity in our experiments.

4.1   Dataset

The fMRI data were obtained from the ADNI initiative [9]. fMRI images of subjects brains were taken every two seconds and are used to compute the Blood-Oxygenation-Level-Dependent (BOLD) signals for different anatomical brain regions. To do this the fMRI voxels were aggregated into larger regions of interest (ROIs). The different ROIs correspond to different anatomical regions of the brain and are assigned anatomical labels to distinguish them. There are 90 such anatomical regions in each fMRI image. The correlation between the average time series in different ROIs represents the degree of functional connectivity between regions which are driven by neural activities.

We construct a graph to represent the pattern of activities using the cross-correlation coefficients for the average time series for pairs of ROIs. We create an undirected edge between two ROI's if the cross-correlation coefficient between the time series is in the top 40% of the cumulative distribution. This cross-correlation threshold is fixed over all of the available data, which provides an optimistic bias for constructing graphs. Those ROIs that have missing time series data are discarded. Subjects fall into different categories according to the degree of severity of the disease, there are normal subjects, those with early mild cognitive impairment, those with late mild cognitive impairment and those with full Alzheimer's. The data supplied 705 patients, including 105 subjects with Alzheimer's disease (AD), 193 normal healthy control subjects (NC), 240 in the Early Mild cognitive impairment (EMCI) and 167 in the Late Mild cognitive impairment (LMCI).

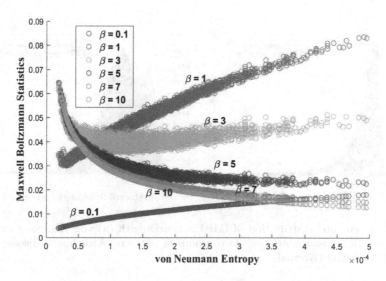

Fig. 1. Scatter plot of thermodynamic edge entropy compared to the von Neumann entropy with different value of temperatures.

4.2   Experimental Results

We first investigate the relationship between the mean edge entropy computed using Maxwell-Boltzmann statistics and von Neumann entropy. Figure 1 shows the edge entropy with varying temperatures. The statistical entropy exhibits a transition in behaviour with respect to the von Neumann entropy with varying temperature.

For example, at the high temperature ($\beta = 0.1$), the thermodynamic entropy is roughly in linear proportion to the von Neumann entropy. As the temperature reduces, they take on an approximately exponential dependence. At low temperature, the thermodynamic edge entropies decrease monotonically with the von Neumann edge entropy ($\beta = 10$). Therefore, at high temperature, the statistical and von Neumann edge entropies are proportional, while at low temperature they are in inverse proportion.

Then, we apply the edge entropy computations to fMRI brain networks, with the aim of determining which anatomical regions play the strongest role in the development of Alzheimer's disease. Figure 2 shows the different edge entropy distribution for the Alzheimer's disease (AD) and healthy control (Normal) samples. Compared to the von Neumann entropy which does not show a clear difference in distributions between the two groups, the thermodynamic entropy better distinguish the detailed distribution of edge entropy. The edge entropy in the case Alzheimer's disease tends towards lower values. This observation is more palpable in the cases of the Maxwell-Boltzmann edge entropy distributions, as shown in Fig. 2(b), with more edges tending to occupy the low entropy region.

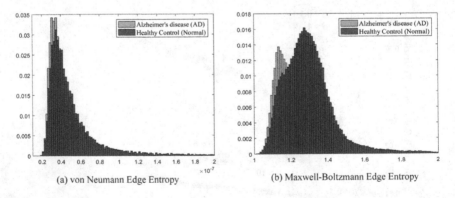

(a) von Neumann Edge Entropy

(b) Maxwell-Boltzmann Edge Entropy

Fig. 2. Edge entropy distribution of fMRI networks with (a) von Neumann entropy, (b) Maxwell-Boltzmann statistics. Two groups of patients, Alzheimer's disease (AD) and healthy control (Normal).

Next, we select the edges with the largest 10% of entropy in the anatomical regions to reduce the feature dimension. This gives 278 significant edges as a feature vector. We explore whether these feature vectors can be used to classify

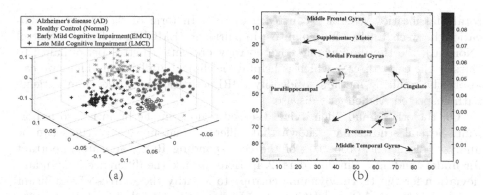

Fig. 3. (a) Visualisation of three dimensional principal components of thermodynamic edge entropy in four groups of Alzheimer's disease. (b) Significant differences between edge entropy associated with diseased areas in the brain. We use the standard deviation of thermodynamic entropy to identify the divergence between AD and HC groups for each edge.

normal healthy subjects and patients with early Alzheimer's disease. Figure 3(a) is the visualisation of the three-dimensional principal components for four groups using Fisher's linear discriminant analysis (LDA). Three principal eigenvectors show the cluster of each group. The palpable feature is that the statistical edge entropy in Maxwell-Boltzmann case can give the separation among the four subject groups.

If we regard the principal components as the feature vectors on each sample graph, we can apply C-SVM (Support Vector Machine) to classify four groups. The data are manually separated into two parts as 500 samples for the training data and 205 samples for the testing data (the rest of the raw data excluding the training set). The 10-fold cross-validation with the grid search method is used to find the optimal parameters (c and g)in C-SVM with Gaussian kernel. The training and testing accuracies are shown in Table 1.

Table 1. SVM Classification Accuracy. In the four group classification, 500 samples are used for training and 205 samples for testing. For AD and Normal binary classification, 200 samples are used for training and 98 samples for testing. For EMCI and Normal binary classification, 300 samples are used for training and 133 samples for testing.

	Training accuracy	Testing accuracy
Four groups	88.42% (442/500)	87.80% (180/205)
AD/Norm	83.50% (167/200)	82.65% (81/98)
EMCI/Norm	92.71% (278/300)	91.03% (121/133)

Table 1 shows that the edge entropies in Maxwell-Boltzmann statistics are good features to identify Alzheimer's disease. For all the groups of patients, the total classification accuracy can reach 87.80%. In term of the binary classification between Early Mild cognitive impairment (EMCI) and healthy control (Normal), the thermodynamic edge entropy presents a better performance to classify the early disease which is helpful for clinical application. Thus, we can apply the resulting method to identify fMRI activation networks from patients with suspected Alzheimer's disease.

On the other hand, identifying diseased regions in the brain is also important. Several studies have shown that different anatomical structures can be analysed using the properties of the corresponding ROIs, and are important for understanding brain disorders [2,4]. Here, we use the difference in standard deviation for the thermodynamic entropy to identify the sources of significant variance between AD and HC groups. Figure 3(b) plots the greatest variance of edge entropy for different anatomical regions (edges). The entropic measurements in the brain areas, such as the Paracingulate Gyrus, Parahippocampal Gyrus, Inferior Temporal Gyrus and Temporal Fusiform Cortex, suggest that subjects with AD experience loss of interconnection between these regions in their brain network during the progression of the disease.

Table 2. Top 10 ROIs with the most significant difference in edge entropy between the Alzheimer's disease (AD) and Health Control (Normal) groups.

Index	ROI	ROI
1	Middle Frontal Gyrus Right(10)	Inferior Parietal Lobule Right(62)
2	Inferior Frontal Gyrus Left(11)	Supramarginal Gyrus Left(63)
3	Supplementary Motor Area Left(19)	Supplementary Motor Area Right(20)
4	Medial Frontal Gyrus Left(23)	Medial Frontal Gyrus Right(24)
5	Posterior Cingulate Gyrus Left(35)	Precuneus Left(67)
6	Hippocampus Left(37)	ParaHippocampal Gyrus Right(40)
7	Hippocampus Right(38)	Amygdala Right(42)
8	ParaHippocampal Gyrus Left(39)	ParaHippocampal Gyrus Right(40)
9	Lingual Gyrus Left(47)	Middle Occipital Gyrus Left(51)
10	Angular Gyrus Left(65)	Angular Gyrus Right(66)

As listed in Table 2, the ten anatomical regions with the largest entropy differences for subjects with the full AD are Paracingulate Gyrus, Parahippocampal Gyrus, Temporal Fusiform Cortex, etc. This result is consistent with the previous study reported in [4,5]. For example, the parahippocampal gyrus has consistently been reported as being vulnerable to pathological changes in Alzheimer's disease (AD), which is closely related to entorhinal and perirhinal subdivisions as the most heavily damaged cortical areas for the disease [13]. The Frontal Medial Cortex and Temporal Fusiform Cortex are memory-related cognitive areas. They are severely damaged by Alzheimer's disease and affect recognition memory for faces. Overall, the loss of connection between these brain regions results in significant functional impairment between healthy subjects and patients with the AD.

In conclusion, both statistical methods and von Neumann edge entropies can be used to represent changes in network structure. Compared to the von Neumann edge entropy, thermodynamic edge entropies are more sensitive to sample variance associated with the degree distribution. Maxwell-Boltzmann statistics reflect strong community structure which is more suitable for representing a detailed structure of the degree distribution.

5   Conclusion

In this paper, we show how to decompose the global network entropies resulting from thermodynamic occupation statistics onto the constituent edges of a graph. We refer to the resulting statistical quantities as Maxwell-Boltzmann edge-entropies. The method uses the normalised Laplacian matrix as the Hamiltonian operator of the network to compute the corresponding partition functions. We undertake experiments to analyse the thermodynamic edge entropies and compare them to their von Neumann counterparts. Experiments reveal that the Maxwell-Boltzmann edge entropy distributions can effectively in characterising detailed variations in the network structure. It outperforms the von Neumann entropy in this respect. Finally, we apply this novel method to provide insights into the neuropathology of Alzheimer's disease. The thermodynamic edge entropy distribution is capable of discriminating between subjects suffering from Alzheimer's and healthy subjects.

References

1. van den Heuvel, M.P., Pol, H.E.H.: Exploring the brain network: a review on resting-state fMRI functional connectivity. J. Eur. Neuropsychopharmacol. **20**, 519–534 (2010)
2. Rubinov, M., Sporns, O.: Complex network measures of brain connectivity: uses and interpretations. Neuroimage **52**(3), 1059–69 (2010)
3. Ye, C., Wilson, R.C., Comin, C.H., Costa, L.D.F., Hancock, E.R.: Approximate von Neumann entropy for directed graphs. Phys. Rev. E **89**(5), 052804 (2014)
4. Rombouts, S.A., Barkhof, F., Goekoop, R., Stam, C.J., Scheltens, P.: Altered resting state networks in mild cognitive impairment and mild Alzheimer's disease: an fMRI study. Hum. Brain Mapp. **26**(4), 231–239 (2005)
5. Khazaee, A., Ebrahimzadeh, A., Babajani-Ferem, A.: Classification of patients with MCI and AD from healthy controls using directed graph measures of resting-state fMRI. Behav. Brain Res. **322**, 339–350 (2016)
6. Wang, J., Wilson, R.C., Hancock, E.R.: Spin statistics, partition functions and network entropy. J. Complex Netw. **5**(6), 858–883 (2017)
7. Passerini, F., Severini, S.: The von Neumann entropy of networks. Int. J. Agent Technol. Syst. **1**, 58–67 (2008)
8. Han, L., Escolano, F., Hancock, E.R., Wilson, R.C.: Graph characterizations from von Neumann entropy. Pattern Recognit. Lett. **33**, 1958–1967 (2012)

9. Alzheimer's Disease Neuroimaging Initiative (ADNI). http://adni.loni.usc.edu/
10. Ye, C., Wilson, R.C., Hancock, E.R.: An entropic edge assortativity measure. In: Liu, C.-L., Luo, B., Kropatsch, W.G., Cheng, J. (eds.) GbRPR 2015. LNCS, vol. 9069, pp. 23–33. Springer, Cham (2015). https://doi.org/10.1007/978-3-319-18224-7_3
11. Wang, J., Wilson, R.C., Hancock, E.R.: Network edge entropy from Maxwell-Boltzmann statistics. In: Battiato, S., Gallo, G., Schettini, R., Stanco, F. (eds.) ICIAP 2017. LNCS, vol. 10484, pp. 254–264. Springer, Cham (2017). https://doi.org/10.1007/978-3-319-68560-1_23
12. Wang, J., Wilson, R.C., Hancock, E.R.: fMRI activation network analysis using Bose-Einstein entropy. In: Robles-Kelly, A., Loog, M., Biggio, B., Escolano, F., Wilson, R. (eds.) S+SSPR 2016. LNCS, vol. 10029, pp. 218–228. Springer, Cham (2016). https://doi.org/10.1007/978-3-319-49055-7_20
13. Van Hoesen, G.W., Augustinack, J.C., Dierking, J., Redman, S.J., Thangavel, R.: The parahippocampal gyrus in Alzheimer's disease: clinical and preclinical neuroanatomical correlates. Ann. New York Acad. Sci. 911(1), 254–274 (2000)

Solving the Graph Edit Distance Problem with Variable Partitioning Local Search

Mostafa Darwiche[1,2]([⊠]), Donatello Conte[1], Romain Raveaux[1],
and Vincent T'kindt[2]

[1] Laboratoire d'Informatique Fondamentale et Appliquée de Tours (EA 6300),
University of Tours, Tours, France
{mostafa.darwiche,donatello.conte,romain.raveaux}@univ-tours.fr
[2] Laboratoire d'Informatique Fondamentale et Appliquée de Tours (EA 6300),
ERL-CNRS 6305, University of Tours, Tours, France
tkindt@univ-tours.fr

Abstract. In the world of graph matching, the *Graph Edit Distance* (GED) problem is a well-known distance measure between graphs. It has been proven to be a $\mathcal{NP}$-hard minimization problem. This paper presents an adapted version of *Variable Partitioning Local Search* (VPLS) matheuristic for solving the GED problem. The main idea in VPLS is to perform local searches in the solution space of a *Mixed Integer Linear Program* (MILP). A local search is done in a small neighborhood defined based on a set of special variables. Those special variables are selected based on a procedure that extracts useful characteristics from the instance at hand. This actually ensures that the neighborhood contains high quality solutions. Finally, the experimentation results have shown that VPLS has outperformed existing heuristics in terms of solution quality on CMU-HOUSE database.

Keywords: Graph Edit Distance · Graph Matching ·
Mixed Integer Linear Program · Variable Partitioning Local Search ·
Matheuristic

1 Introduction

Graphs are heavily involved in *Structural Pattern Recognition* (SPR). Using graphs, it is possible to model objects and patterns by considering the main components as vertices and expressing the relations between those components using edges. Moreover, graphs can store extra properties and characteristics about the pattern by assigning attributes to vertices and edges. Then, these graphs are exploited to perform object comparison and recognition [15]. In fact, this is known as *Graph Matching* (GM), which is the core of the SPR field. GM is about finding vertices and edges mappings between two graphs, from which a (dis-)similarity measure can be computed. In addition, GM covers many problems that are split into two main categories: *Exact* (EGM) and *Error-Tolerant* (ETGM). The main

© Springer Nature Switzerland AG 2019
D. Conte et al. (Eds.): GbRPR 2019, LNCS 11510, pp. 67–77, 2019.
https://doi.org/10.1007/978-3-030-20081-7_7

difference between the two categories is that EGM requires having the same topologies and attributes in graphs. While ETGM is flexible and accommodates to differences in graphs. ETGM is more preferable because it is unlikely to have the same exact graphs in real-life scenarios. Among the various problems that fall into ETGM category, the *Graph Edit Distance* (GED) problem is considered as the most popular one. Solving this problem computes a dissimilarity measure between two graphs [4]. The main idea in GED is to transform one source graph into another target graph, by applying a certain number of edit operations. Those edit operations are: substitution, insertion and deletion of a vertex/edge. Each edit operation has an associated cost. The aim when solving the GED problem is to find a set of edit operations that minimizes the total cost. This is what makes it a very complex problem to solve, which later was proven to be a $\mathcal{NP}$-hard minimization problem [19]. Despite the complex nature of the problem, it is still seen as an important one because it was shown to be a generalization to other GM problems such as the maximum common subgraph and the subgraph isomorphism [2,3]. Also, the GED has applications in many fields such as image analysis, biometrics, bio/cheminformatics, etc. [18].

Looking into the literature, numerous methods for solving the GED problem exist. They can be divided into two classes: exact and heuristics. In the first class, there are methods that solve an instance to the problem to optimality. Such methods tend to become expensive when dealing with large graphs, because of the exponential growth in complexity. The other class, however, contains heuristic methods that aim at computing a sub-optimal solution in a reasonable amount of time. For exact methods, mathematical programming is used to model the GED problem providing *Mixed Integer Linear Models* (MILP). Two formulations appear to outperform other methods: JH by Justice and Hero [11] and F3 by Darwiche et al. [8]. JH is designed to solve a sub-problem of the GED (denoted by GED^{EnA}), in which the attributes on edges are ignored. However, F3 is designed to solve instances of the general GED problem. Regarding the heuristic methods, there are plenty of them. Starting with the most famous and fastest one the Bipartite GM heuristic (denoted shortly by BP), which was developed by Riesen et al. [16]. BP breaks down the GED problem into a linear sum assignment problem that can be solved in polynomial time, using the Hungarian algorithm [14]. Later, BP has been improved in many works such as *FastBP* and *SquareBP* [17], and also it has been used in other heuristics such as *SBPBeam* [10]. Other heuristics are, for instance, *Integer Projected Fixed Point* (IPFP) and *Graduate Non Convexity and Concavity Procedure* (GNCCP) [1]. They are based on solving the *Quadratic Assignment Problem* (QAP) model for the GED problem proposed by the same authors. A recent heuristic method has been designed in [7] and referred to as *LocBra*. It is based on local searches in the solution space of a MILP formulation. This kind of heuristics on the basis of MILP formulations is known by *Matheuristics*. LocBra was shown in [6,7] to be more effective than existing heuristics (e.g. SBPBeam, IPFP and GNCCP) when dealing with instances of GED^{EnA}.

This work is an attempt to design a new matheuristic that can accomplish accurate results as LocBra but on the general problem. It is actually an adapted version of *Variable Partitioning Local Search* (VPLS) matheuristic proposed originally by Della Croce et al. [9]. The main ingredients in VPLS are: a MILP formulation which is going to be *F*3 and a MILP solver which is CPLEX. Then, VPLS defines neighborhoods around feasible solutions by modifying the MILP. The modified formulation will be handed over to the solver to explore the neighborhoods looking for improved solutions. The special version dedicated to the GED problem involves integrating problem-dependent information and characteristics into the neighborhood definition, which increases most likely the performance of the heuristic.

The remainder is organized as follows: Sect. 2 presents the definition of GED problem, followed with a review of *F*3 formulation. Then, Sect. 3 details the proposed heuristic. And Sect. 4 shows the results of the computational experiments. Finally, Sect. 5 highlights some concluding remarks.

2   *GED* Definition and *F*3 Formulation

2.1   *GED* Problem Definition

Given two graphs $G = (V, E, \mu, \xi)$ and $G' = (V', E', \mu', \xi')$, GED is the task of transforming the graph source G into the graph target G'. To accomplish this, GED introduces vertex and edge edit operations: $(i \rightarrow k)$ is the substitution of two vertices, $(i \rightarrow \epsilon)$ is the deletion of a vertex, and $(\epsilon \rightarrow k)$ is the insertion of a vertex, with $i \in V, k \in V'$ and ϵ refers to the empty node. The same logic goes for edges. The set of operations that reflects a valid transformation of G into G' is called a *complete edit path*, defined as $\lambda(G, G') = \{o_1, ..., o_k\}$, where o_i is an elementary vertex (or edge) edit operation and k is the number of operations. GED is then,

$$d_{min}(G, G') = \min_{\lambda \in \Gamma(G, G')} \sum_{o_i \in \lambda} \ell(o_i) \qquad (1)$$

where $\Gamma(G, G')$ is the set of all complete edit paths between G and G', d_{min} represents the minimal cost obtained by a complete edit path $\lambda(G, G')$, and $\ell(.)$ is a cost function that assigns costs to elementary edit operations.

2.2   Mixed Integer Linear Program

The general MILP formulation is of the form:

$$\min_{x} c^T x \qquad (2)$$

$$Ax \geq b \qquad (3)$$

$$x_i \in \{0, 1\}, \forall i \in B \qquad (4)$$

$$x_j \in \mathbb{N}, \forall j \in I \qquad (5)$$

$$x_k \in \mathbb{R}, \forall k \in C \tag{6}$$

where $c \in \mathbb{R}^n$ and $b \in \mathbb{R}^m$ are vectors of coefficients, $A \in \mathbb{R}^{m \times n}$ is a matrix of coefficients. x is a vector of variables to be computed. The variable index set is split into three sets (B, I, C), which stand for binary, integer and continuous, respectively. This formulation minimizes an objective function (Eq. 2) w.r.t. a set of linear inequality constraints (Eq. 3) and the bounds imposed on variables x e.g. integer or binary (Eqs. 4–6). A feasible solution is a vector x with the proper values based on their defined types, that satisfies all the constraints. The optimal solution is a feasible solution that has the minimum objective function value. This approach of modeling decision problems (i.e. problems with binary and integer variables) is very efficient, especially for hard optimization problems.

2.3 $F3$ Formulation

$F3$ is a recent MILP formulation proposed by Darwiche et al. [8], which was an improvement to an earlier version (referred to as $F2$) designed by Lerouge et al. [12]. $F3$ is a compact formulation with a set of constraints independent from the edges in the graphs. For this reason, F3 is more effective than $F2$ especially in the case of dense graphs [8]. In the following, $F3$ is detailed by defining: data, variables, objective function to minimize and constraints to satisfy.

Data. Given two graphs $G = (V, E, \mu, \xi)$ and $G' = (V', E', \mu', \xi')$, the cost functions, in order to compute the cost of each vertex/edge edit operations, are known and defined. Therefore, vertices cost matrix $[c_v]$ is computed as in Eq. 7 for every couple $(i, k) \in V \times V'$. The ϵ column is added to store the cost of deletion of i vertices, while the ϵ row stores the costs of insertion of k vertices. Following the same process, the matrix $[c_e]$ is computed for every $((i, j), (k, l)) \in E \times E'$, plus the row/column ϵ for deletion and insertion of edges.

$$c_v = \begin{bmatrix} c_{1,1} & c_{1,2} & \cdots & c_{1,|V'|} & c_{1,\epsilon} \\ c_{2,1} & c_{2,2} & \cdots & c_{2,|V'|} & c_{2,\epsilon} \\ \vdots & \vdots & \ddots & \vdots & \vdots \\ c_{|V|,1} & c_{|V|,2} & \cdots & c_{|V|,|V'|} & c_{|V|,\epsilon} \\ c_{\epsilon,1} & c_{\epsilon,2} & \cdots & c_{\epsilon,|V|} & 0 \end{bmatrix} \begin{matrix} u_1 \\ u_2 \\ \vdots \\ u_{|V|} \\ \epsilon \end{matrix} \tag{7}$$

$$\begin{matrix} v_1 & v_2 & \cdots & v_{|V'|} & \epsilon \end{matrix}$$

Variables. $F3$ focuses on finding the correspondences between the sets of vertices and the sets of edges. So, two sets of decision variables are needed.

- $x_{i,k} \in \{0, 1\}$ $\forall i \in V, \forall k \in V'$; $x_{i,k} = 1$ when vertices i and k are matched, and 0 otherwise.
- $y_{ij,kl} \in \{0, 1\}$ $\forall (i, j) \in E, \forall (k, l) \in E' \cup \overline{E}'$ such that $\overline{E}' = \{(l, k) : \forall (k, l) \in E'\}$; $y_{ij,kl} = 1$ when edge (i, j) is matched with (k, l), and 0 otherwise.

Objective Function. The objective function to minimize is the following.

$$\min_{x,y} \sum_{i \in V} \sum_{k \in V'} (c_v(i,k) - c_v(i,\epsilon) - c_v(\epsilon,k)) \cdot x_{i,k} \tag{8}$$

$$+ \sum_{(i,j) \in E} \sum_{(k,l) \in E'} (c_e(ij,kl) - c_e(ij,\epsilon) - c_e(\epsilon,kl)) \cdot y_{ij,kl} + \gamma$$

The objective function minimizes the cost of assigning vertices and edges with the cost of substitution subtracting the cost of insertion and deletion. The γ, which is a constant given in Eq. 9, compensates the subtracted costs of the assigned vertices and edges. This constant does not impact the optimization algorithm and it could be removed. It is there to obtain the GED value. So at first, the function considers all vertices and edges of G as deleted and the ones of G' as inserted. Then, it solves the problem of finding the cheapest substitution assignments of vertices and edges.

$$\gamma = \sum_{i \in V} c_v(i,\epsilon) + \sum_{k \in V'} c_v(\epsilon,k) + \sum_{(i,j) \in E} c_e(ij,\epsilon) + \sum_{(k,l) \in E'} c_e(\epsilon,kl) \tag{9}$$

Constraints. $F3$ has 3 sets of constraints.

$$\sum_{k \in V'} x_{i,k} \leq 1 \; \forall i \in V \tag{10}$$

$$\sum_{i \in V} x_{i,k} \leq 1 \; \forall k \in V' \tag{11}$$

$$\sum_{(i,j) \in E} \sum_{(k,l) \in E' \cup \overline{E}'} y_{ij,kl} \leq d_{i,k} \times x_{i,k} \; \forall i \in V, \forall k \in V'$$

$$\text{with } d_{i,k} = \min\left(degree(i), degree(k)\right) \tag{12}$$

Constraints 10 and 11 are to make sure that a vertex can be only matched with maximum one vertex. Next, constraints 12 guarantee preserving edges matching between two couple of vertices.

3 VPLS Heuristic

3.1 Main Features of VPLS

Variable Partitioning Local Search (VPLS) is a matheuristic proposed by Della Corce et al. [9]. It aims at solving optimization problems by embedding a MILP solver into heuristic algorithms. More generally, VPLS is about performing neighborhood exploration in the solution space of a MILP formulation. To start a VPLS heuristic, two ingredients are needed: MILP formulation and MILP solver.

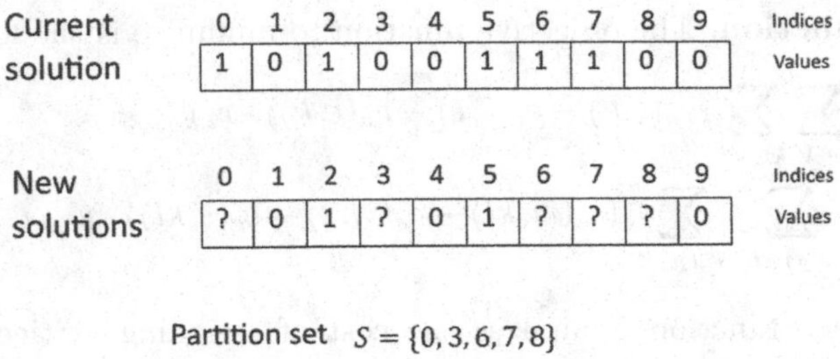

| Current | 0 | 1 | 2 | 3 | 4 | 5 | 6 | 7 | 8 | 9 | Indices |
| solution | 1 | 0 | 1 | 0 | 0 | 1 | 1 | 1 | 0 | 0 | Values |

| New | 0 | 1 | 2 | 3 | 4 | 5 | 6 | 7 | 8 | 9 | Indices |
| solutions | ? | 0 | 1 | ? | 0 | 1 | ? | ? | ? | 0 | Values |

Partition set $S = \{0, 3, 6, 7, 8\}$

Fig. 1. Example of VPLS partitioning.

The first step in VPLS heuristic is to compute a feasible solution $\bar{X}$. Let $X_B = \{x_i | \forall i \in B\}$ be the set of binary variables and $\bar{X}_B = \{\bar{x}_i | \forall i \in B\}$ be the set of values assigned to binary variables based on $\bar{X}$. Now, assuming that there exists a partition $S \subseteq B$ of "special" binary variables. The variables in S are selected based on some defined rules, where these rules underlie some analyses and observations related to the problem. Later, a procedure is presented for selecting those special variables based on problem-dependent information and characteristics of an instance. After determining the set S, a neighborhood $N(\bar{X}, S)$ can be defined as follows:

$$N(\bar{X}, S) = \{X_B \mid x_j = \bar{x}_j, \forall j \notin S\} \tag{13}$$

The neighborhood of $\bar{X}$, then, contains all solutions of the MILP such that, they share the same values of binary variables not belonging to subset S, as in the current solution $\bar{X}_B$. Meanwhile, the variables belonging to subset S remain free. An example of variables partitioning is depicted in Fig. 1. So, the resulting restricted MILP formulation has a part of its binary variables with default values (as in the solution $\bar{X}$). At this point, the solver can be called to solve the restricted formulation looking for the optimal/best solution in the neighborhood $N(\bar{X}, S)$. The new solution is the optimal in that neighborhood, if the proof of optimality is returned by the solver. In the case where the restricted formulation is difficult, then the solver can be forced to stop and return the best solution found so far. This step stands for the search intensification in VPLS. Finally, the current solution $\bar{X}$ is updated with the new solution. To sum up, VPLS consists of three main steps:

1. Neighborhood definition around a current solution $\bar{X}$.
2. Intensifying the search in the neighborhood.
3. Updating the current solution with the new one.

The process can be repeated until a defined stopping criterion is met.

3.2   VPLS for the GED Problem

To make the heuristic suitable for the GED problem, $F3$ is selected as the main formulation. Then, A fundamental question arises when implementing VPLS is how to define the set S? Earlier, the variables in S were referred to as special variables, and this is to indicate that they should be chosen carefully. Choosing them randomly is a possibility, but there is no guarantee that the neighborhoods will contain good and diversified solutions.

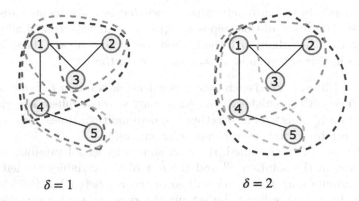

$$\delta = 1 \qquad\qquad\qquad \delta = 2$$

Fig. 2. Example of generating spheres for a graph. When $\delta = 1$, in red is the sphere for vertex 1, in green is the sphere for vertices 2 and 3, in orange is the sphere for vertex 4 and in blue is the sphere for vertex 5. When $\delta = 2$, in red is the sphere for vertices 1 and 4, in green is the sphere for vertices 2, 3, and in blue is the sphere for vertex 5. (Color figure online)

So, back to defining the set S, it is essential to select variables that affect the most the matching (and at the same time the objective function). Basically, only $x_{i,k}$ variables are going to be considered when defining the set S. And next, a procedure based on the notion of spheres is followed to determine S. This procedure needs two input graphs G and G' and an initial solution x^0, and it proceeds as follows:

(i) First, define the list of spheres on graph G of radius δ. For each vertex i in G, the sphere $\mathscr{S}_i$ contains all vertices j that are distant from i with at most δ edges, e.g. if $\delta = 1$, $\mathscr{S}_i$ contains all vertices connected to i with an edge. To compute how many edges are needed to go from one vertex to another, the well-known Dijkstra algorithm is used [5]. It computes the shortest path between two vertices in a graph. In fact, each sphere is a subgraph of G, containing all vertices accessible by at most δ edges, plus the edges connecting any two vertices in the sphere. Figure 2 shows an example of spheres with different δ values.

(ii) Next, compute a cost for each sphere $\mathscr{S}_i$ based on the assignments in the initial solution x^0. For example, if $\mathscr{S}_1$ for vertex u_1 contains vertices $\{u_1, u_2, u_3, u_4\}$. From the solution x^0, see to which vertex $k \in V'$ the vertex u_1 is assigned, and include the cost $c(u_1 \to v)$ to the sphere's cost. As well, check the edges that are part of sphere $\mathscr{S}_1$ and find their assignments so their costs are added to the sphere's cost. The same is done for the rest of the vertices ($\{u_2, u_3, u_4\}$).

$$c_{\mathscr{S}} = \sum_{\forall i \in \mathscr{S}} c(i \to assign(i)) + \sum_{\forall (i,j) \in (\mathscr{S} \times \mathscr{S}) \cap E} c((i,j) \to assign((i,j))), \quad (14)$$

with *assign* a function to determine vertices/edges assignments based on x^p solution. The result of this step is an array $[c_{\mathscr{S}}]$ storing costs of all spheres.

(iii) Finally, find the sphere with the highest cost in $[c_{\mathscr{S}}]$ array. Then, for every vertex i in this sphere, add all $x_{i,k}$ variables to the set S.

Steps (ii) and (iii) are called each time a new feasible solution is found to select the next sphere with the highest cost. An already selected sphere is excluded in the next iteration. This avoids selecting a sphere multiple times, and searching in the same neighborhood several times consecutively.

Once the set S is determined, the next step is to set all variables not in S to their values in the solution x^0 and the rest of the variables are left free in the MILP formulation. The solver will solve the restricted MILP formulation trying to find the best solution by setting the right values for variables in S. This is the intensification phase, that will result in a new solution x^1. Again, the spheres costs are recomputed based on x^1, and the one with the highest cost will be selected for the next iteration. An iteration, then, consists of three steps: computing and selecting the sphere to define S, defining the neighborhood based on S, intensifying the search in the neighborhood. This process is repeated until reaching some defined stopping criterion.

Finally, VPLS requires the following parameters to be set:

1. δ, is the radius of spheres.
2. *total_time_limit*, is the total running time allowed for VPLS before stopping.
3. *node_time_limit*, is the maximum running time given to the solver to solve the restricted MILP formulation.
4. *UB_time_limit*, is the running time allowed to the solver to compute an initial solution.
5. *cons_sol_max*, serves as a stopping criterion: VPLS stops when the number of consecutive intensification steps finding solutions with the same objective function values is equal to this parameter.

4   Computational Experiments

Database. Among the numerous existing databases, CMU-HOUSE database is selected in this experiment [13]. It contains 111 (attributed and undirected) graphs corresponding to $3D$-images of houses. The particularity of this database is that graphs are extracted from $3D$-images of houses, where the houses are rotated with different angles. This is interesting because it enables testing and comparing graphs representing the same house but positioned differently inside the images. The total number of instances is 660.

Experiment Settings and Comparative Heuristics. VPLS algorithm is implemented in C language. The solver CPLEX 12.7.1 is used to solve the MILP formulations. CPLEX solver is configured to use a single thread, and the rest of the parameters are set to default. Experiments are ran on a machine with Windows 7×64, Intel Xeon $E5$ 2.30 GHz, 4 cores and 8 Gb of RAM. The following heuristics are selected from the literature in the comparison: SBPBeam [10], IPFP and GNCCP [1]. Their parameters are set to the values as mentioned in the references. VPLS parameters are set empirically to the following values: $\delta = 2$, $cons\_sol\_max = 5$, $total\_time\_limit = 10\,s$, $node\_time\_limit = 2\,s$, $UB\_time\_limit = 4\,s$. For each heuristic, the following indicators are computed: t_{min}, t_{avg}, and t_{max} are the minimum, average and maximum CPU times in seconds over all instances. Correspondingly, d_{min}, d_{avg}, and d_{max} are the deviations of the solutions obtained by one heuristic, from the best solutions found by all heuristics (i.e. given an instance I and a heuristic H, deviation percentage is equal to $\frac{solution_I^H - best_I}{best_I} \times 100$, with $best_I$ the best solution for I found by all heuristics).

Table 1. VPLS vs. heuristics on CMU-HOUSE instances

	VPLS	SBPBeam	IPFP	GNCCP
t_{min}	5.76	7.54	0.03	6.85
t_{avg}	9.24	8.50	**0.18**	9.61
t_{max}	10.03	9.72	0.32	11.70
d_{min}	0.00	0.00	0.00	0.00
d_{avg}	**13.13**	330.24	313.05	336.81
d_{max}	294.41	5502.39	34308.00	32426.89
η_I	**537**	126	310	440

Results and Analysis : Based on the results reported in Table 1, VPLS seems to be the best heuristic in terms of solutions quality, with the best average deviation at 13% and η_I at 537. The difference is remarkably high (around 300%) compared to the deviations obtained by other heuristics. Even when looking at worst deviations the difference is very high. However, in terms of average running time, the fastest heuristic is IPFP with t_{avg} at 0.18 s, while other heuristics including VPLS reaches 9 s. Eventually, VPLS has been able to outperform the existing heuristics by obtaining very good solutions.

5   Conclusion

In this work, a VPLS heuristic is designed for solving the GED problem. This heuristic is based on performing local searches on the basis of a MILP formulation. To perform local searches, the neighborhoods are defined based on special variables determined by extracting characteristics from the instance at hand. By doing so, the performance of VPLS improves, which was shown in the experiments where VPLS outperformed the existing heuristics in terms of solution quality. This is a second matheuristic designated to solve the GED problem after the first and successful attempt with local branching [7]. Indeed, matheuristics are effective and are new ways for solving the GED problem. Meanwhile, it will be interesting to combine VPLS and local branching into one matheuristic by unifying the neighborhood definitions. As well, more evaluations and experiments need to be performed to test the methods on different kinds of graphs.

References

1. Bougleux, S., Brun, L., Carletti, V., Foggia, P., Gaüzère, B., Vento, M.: Graph edit distance as a quadratic assignment problem. Pattern Recogn. Lett. **87**, 38–46 (2017)
2. Bunke, H.: On a relation between graph edit distance and maximum common subgraph. Pattern Recogn. Lett. **18**(8), 689–694 (1997)
3. Bunke, H.: Error correcting graph matching: on the influence of the underlying cost function. IEEE Trans. Pattern Anal. Mach. Intell. **21**(9), 917–922 (1999)
4. Bunke, H., Allermann, G.: Inexact graph matching for structural pattern recognition. Pattern Recogn. Lett. **1**(4), 245–253 (1983)
5. Cormen, T.H.: Section 24.3: Dijkstra's algorithm. In: Introduction to Algorithms, pp. 595–601 (2001)
6. Darwiche, M., Conte, D., Raveaux, R., T'Kindt, V.: Graph edit distance: accuracy of local branching from an application point of view. Pattern Recogn. Lett. (2018). https://doi.org/10.1016/j.patrec.2018.03.033. http://www.sciencedirect.com/science/article/pii/S0167865518301119
7. Darwiche, M., Conte, D., Raveaux, R., T'Kindt, V.: A local branching heuristic for solving a graph edit distance problem. Comput. Oper. Res. (2018). https://doi.org/10.1016/j.cor.2018.02.002. http://www.sciencedirect.com/science/article/pii/S0305054818300339
8. Darwiche, M., Raveaux, R., Conte, D., T'Kindt, V.: Graph edit distance in the exact context. In: Bai, X., Hancock, E.R., Ho, T.K., Wilson, R.C., Biggio, B., Robles-Kelly, A. (eds.) S+SSPR 2018. LNCS, vol. 11004, pp. 304–314. Springer, Cham (2018). https://doi.org/10.1007/978-3-319-97785-0_29
9. Croce, F.D., Grosso, A., Salassa, F.: Matheuristics: embedding MILP solvers into heuristic algorithms for combinatorial optimization problems. In: Siarry, P. (ed.) The Oxford Handbook of Innovation, Chap. 3. NOVA Publisher (2013)
10. Ferrer, M., Serratosa, F., Riesen, K.: Improving bipartite graph matching by assessing the assignment confidence. Pattern Recogn. Lett. **65**, 29–36 (2015)
11. Justice, D., Hero, A.: A binary linear programming formulation of the graph edit distance. IEEE Trans. Pattern Anal. Mach. Intell. **28**(8), 1200–1214 (2006)
12. Lerouge, J., Abu-Aisheh, Z., Raveaux, R., Héroux, P., Adam, S.: New binary linear programming formulation to compute the graph edit distance. Pattern Recogn. **72**, 254–265 (2017)
13. Moreno-García, C.F., Cortés, X., Serratosa, F.: A graph repository for learning error-tolerant graph matching. In: Robles-Kelly, A., Loog, M., Biggio, B., Escolano, F., Wilson, R. (eds.) S+SSPR 2016. LNCS, vol. 10029, pp. 519–529. Springer, Cham (2016). https://doi.org/10.1007/978-3-319-49055-7_46

14. Munkres, J.: Algorithms for the assignment and transportation problems. J. Soc. Ind. Appl. Math. **5**(1), 32–38 (1957)
15. Riesen, K.: Structural Pattern Recognition with Graph Edit Distance. Advances in Computer Vision and Pattern Recognition. Springer, Cham (2015). https://doi. org/10.1007/978-3-319-27252-8
16. Riesen, K., Neuhaus, M., Bunke, H.: Bipartite graph matching for computing the edit distance of graphs. In: Escolano, F., Vento, M. (eds.) GbRPR 2007. LNCS, vol. 4538, pp. 1–12. Springer, Heidelberg (2007). https://doi.org/10.1007/978-3-540-72903-7_1
17. Serratosa, F.: Computation of graph edit distance: reasoning about optimality and speed-up. Image Vis. Comput. **40**, 38–48 (2015)
18. Stauffer, M., Tschachtli, T., Fischer, A., Riesen, K.: A survey on applications of bipartite graph edit distance. In: Foggia, P., Liu, C.-L., Vento, M. (eds.) GbRPR 2017. LNCS, vol. 10310, pp. 242–252. Springer, Cham (2017). https://doi.org/10. 1007/978-3-319-58961-9_22
19. Zeng, Z., Tung, A.K., Wang, J., Feng, J., Zhou, L.: Comparing stars: on approximating graph edit distance. Proc. VLDB Endow. **2**(1), 25–36 (2009)

A Database and Evaluation
for Classification of RNA Molecules
Using Graph Methods

Enes Algul[✉] and Richard C. Wilson

University of York, York, UK
{ea918,richard.wilson}@york.ac.uk

Abstract. In this paper, we introduce a new graph dataset based on the representation of RNA. The RNA dataset includes 3178 RNA chains which are labelled in 8 classes according to their reported biological functions. The goal of this database is to provide a platform for investigating the classification of RNA using graph-based methods. The molecules are represented by graphs representing the sequence and base-pairs of the RNA, with a number of labelling schemes using base labels and local shape. We report the results of a number of state-of-the-art graph based methods on this dataset as a baseline comparison and investigate how these methods can be used to categorise RNA molecules on their type and functions. The methods applied are Weisfeiler Lehman and optimal assignment kernels, shortest paths kernel and the all paths and cycle methods. We also compare to the standard Needleman-Wunsch algorithm used in bioinformatics for DNA and RNA comparison, and demonstrate the superiority of graph kernels even on a string representation. The highest classification rate is obtained by the WL-OA algorithm using base labels and base-pair connections.

1 Introduction

Ribonucleic acid (RNA) is chemically very similar to DNA in their polymer of nucleotides [1]. These nucleotides have sequences that can encode genetic information [2]. DNA stores genetic information while RNA, copied from DNA, carries and provides this genetic information to other biological process. RNAs are also well known to play important regulatory and catalytic roles [3]. These roles including transcriptional regulation, RNA splicing, and RNA modification and maturation [3]. RNA also very important for treatment of diseases including viral and bacterial infections, and cancer [4]. RNA is therefore crucial to all life and it is important to understand its function.

The primary structure of the RNA consists of nucleotide sequences, this nucleotide sequences can fold onto itself to create secondary and tertiary structure of the RNA. Unlike DNA, RNA is single strand and is encouraged to fold into complex shapes, like proteins, by the matching of base-pairs from the same strand. The secondary structure is formed by both Watson-Crick base pairs [5],

© Springer Nature Switzerland AG 2019
D. Conte et al. (Eds.): GbRPR 2019, LNCS 11510, pp. 78–87, 2019.
https://doi.org/10.1007/978-3-030-20081-7_8

(A-U, C-G) and non-standard pairs. The base pairs between A-U, C-G, and the wobble pair between G-U are referred to as canonical base pairs while base pairs between other base pairs are called non-canonical base pairs [6]. The canonical base pairs are more stable and important than non-canonical bases in the structure of the RNA [6]. The secondary structures is the topology of the RNA folding, and it consist of five main structural components: called internal loops, hairpins, bulges, junctions, and stems. The geometric shape of the RNA is its tertiary (3D) structure.

The objective of this paper is to present a new, large, graph RNA dataset which can be used to investigate graph-based methods for RNA classification and discovery. We also investigate the performance of some standard methods on the dataset and the role of different elements of the RNA representation, particularly the labelling, topology and geometry of the RNA.

Outline of the Paper: In Sect. 2 we will explain related works. In Sect. 3 we will demonstrate our dataset. In Sect. 4 we will represent RNA molecules. In Sect. 5 we will explain sequence alignment, Weisfeiler-Lehman optimal assignment kernel, all paths and cycle embedding, and shortest path embedding methods. In Sect. 6 we will show our experimental results. In Sect. 7 we will discuss on our experiments and conclude the paper.

2   Related Work

DNA and RNA have chemical and structural similarities. Both molecules are nucleic acid composed of nucleobases, although the sugar backbone of the polymer is different. The structure of the molecule is determined by the nucleotide sequence. Because of this, sequence alignment is commonly used to determine the biological function of the DNA, such as the Needleman Wunch algorithm [7]. This is essentially a string edit distance between the strings of base-labels. The nucleotide sequence is the *primary structure* of the RNA. Because RNA is single strand, it can fold on itself, and the folds can be held in place by base-pair bonding between bases at different points on the RNA strand. This creates the potential for a more complex topology than RNA. This structure is called the *secondary structure* and can naturally be represented by a (labelled) graph.

Graph Theory is a branch of mathematics which has been used in various areas, such as road systems, neurosciences, irrigation networks, chemical processes and structures, computer science, and bioinformatics [1]. Graph-based data is becoming more abundant in chemical pathways and protein structures, protein or gene regulation networks, and social networks [8]. Graph Kernels allow the application of kernel methods to graph data [10] and allow using a range of algorithms for pattern recognition [9]. Graph kernels bridge the gap between graph-structured data and a large spectrum of the machine learning algorithms such as SVM kernel regression, kernel PCA [8], KNN and ensemble classifiers (Subspace KNN, Subspace Discriminant, Bagged Trees, and Boosted Trees). In this work, the goal of applying graph kernels is to measure similarities between two patterns, while the goal of the Machine learning is to classify these similarities. Kernel methods are widely used in the field of the bioinformatics, such as in Lodhi and Huma [10] where the spectrum kernel, marginalise kernel and fisher kernel were applied for sequence analysis.

3 Database

There exist large databases of DNA, RNA and protein structures. In the reviewed literature, most of the dataset are in fasta and protein data bank (pdb) file formats. The fasta files include the basic sequence of macro-molecules (protein, DNA, RNA) [11,12]. The pdb files include the information of each atom of the macro-molecules, their sequence information, and their atomic coordinates. The data which was used in recent studies are in the pdb format which is standardised according to the atomic coordinates [13]. The pdb files provided by the many organisations. The three largest of them are the Protein Data Bank Japan (PDBj) [14], Nucleic Acid Database (NDB) [15], and The Research Collaboratory for Structural Bioinformatics Protein Data Bank (RCSB PDB) [16].

The pdb files consist of more than one chain of information of the RNA molecules. For instance, in the same pdb files there might be different kinds of chains belong to a different type of RNA or DNA or RNA-Protein interaction. For instance, in 1XNR.pdb file, chain A is 16S ribosomal RNA, chain X is anticodon TRNA, chain M is MRNA, and other chains such as chain B, C, D, ..., T, V are 16S ribosomal protein. It is not straightforward to extract single RNA chains from this type of data. The largest classified database of RNA structure (i.e. RNA strands with functional labels) known to us is that of Klosterman et al. [17] which contains 419 RNA strands.

RNA Bricks [18] downloaded pdb files from the World Wide Protein Data Bank web site and split each pdb files by their chains. Their dataset is publicly available. We have extracted 3178 RNA chains from this dataset. For each of this molecules, we have investigated the literature to classify them into one of 8 possible RNA classes. The first step is to look at the MOL-ID field in the pdb files which include information of the type of RNA molecules. In some pdb files, the type of the RNA is not very clear from this field. For these, we undertook a more elaborate investigation using information derived from the HEADER, the TITTLE, the KEYWDS, and HD-RNA [19] in order to determine the type of the RNA chains. We then removed any chains where we were still unsure of the type. The result is a curated database of 3178 RNA molecules with 8 possible class labels which is available for download [20].

The RNA classes from the Table 1, ribozyme is a type of the RNA which catalyses chemical reactions, riboswitches behave like ribozymes and participate gene regulation, ribonucleases are very important enzymes in RNA degradation and maturation pathways, signal recognition particle (SRP) RNA, a part of ribonucleo-protein (protein-RNA complex), involves in targeting translocation of membrane proteins and secretory proteins. We labelled all other RNA classes, which the number of classes too small, in the OTHER section. We did not labelled our dataset according to the source of the organism.

Table 1. The labelled classification of the RNAs and their descriptions. The number of each type of RNAs represents in the brackets.

Classes	Keywords/Description
RIBONUCLEASE (14)	RIBONUCLEASE P, RNASE P
RIBOSWITCH (227)	APTAMER
MRNA (179)	UTR, EXON
RIBOZYME (259)	S-TURN, CATALYTIC RNA, HAMMERHEAD, GLMS
RRNA (1135)	4.8S, 5S, 5.8S,16S, 18S, 23S, 25S, 26S, 28S, 30S, 40S, 50S, 60S, 70S, 80S, A-SITE OF HUMAN RIBOSOME, A-SITE OF HUMAN MITOCHONDRIAL RIBOSOME, A-SITE OF BACTERIAL RIBOSOME, SARCIN/RICIN 28S RRNA
SRP (57)	4.5S, 7S, 7SL
TRNA (581)	A-site, P-site, tRNA X-MER, FMET, FME, INITIATOR, INI, PRIMER CODON, ANTICODON, ACCEPTOR, tRNA-fragment
OTHER (726)	viral RNA, miRNA, snoRNA, IRES RNA, and some undefined RNAs such as 5' RNA, 16-MER, 192-MER, 28-MER, 119-MER, 97-MER etc.

In our data set, 332 RNA chains' nucleotide lengths are from 6 to 9, 1798 RNA chain's nucleotide sizes are from 10 to 100, 469 RNA chains nucleotide sizes are from 101 to 500, 277 RNA chain's nucleotide sizes are from 1326 to 1861, 286 RNA chain's nucleotide sizes are from 2227 to 2912, 15 RNA chain's nucleotide sizes are from 3119 to 3662, and nucleotide size of one chain is 4298. The percentage of the each type of the RNA molecules are 35.71% (RRNA), 22.84% (OTHER), 18.28% (TRNA), 8.15% (RIBOZYME), 7.14% (RIBOSWITCH), 5.63% (MRNA), 1.79% (SRP), and 0.44% (RIBONUCLEASE).

4 RNA Representation

The main component of Nucleic Acid is 5 carbon sugar (2-deoxyribose or ribose), a phosphate group and a base (one of four molecules = adenine, guanine, cytosine, and uracil/thymine). There is two macro nucleic acids called as deoxyribonucleic acid (DNA) and ribonucleic acid (RNA). Ribonucleic Acid (RNA) consist of nucleotides. The nucleotides are composed of purine nucleobases, namely Adenine (A), Guanine (G), and pyrimidine nucleobases calling Cytosine (C) and Uracil (U).

The base sequences represents the primary structure of the RNA such as the following sequence is single stranded (strand B) of an Escherichia coli Riboswitch '4Y1M':

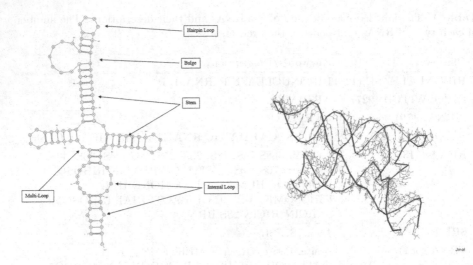

Fig. 1. 3D and secondary structure of the Escherichia coli Riboswitch derived from 4Y1M.pdb. The secondary structure's image produced using RNApdbee 2.0 [21] and 3D structure's image produced using Matlab molviewer.

"GAUUUGGGGAGUAGCCGAUUUCCGAAAGGAAAUGUACGUGUCAA CAUUUUCGUUGAAAAACGUGGCACGUACGGACUGAAGAAAUUCAGU CAGGCGAGACCAUAUCC"

The primary structure of the RNA folds on to itself to build secondary structure and 3D structure of the RNA as represented in Fig. 1. 2D structure of the RNA is widely used to classify and define RNA molecules with their functions. The main components of the 2D RNAs are hairpin loop, bulge, internal loop, multi-loop, and stem. As represented on Fig. 1; the uninterrupted base-paired portion of the RNA molecules is called a stem, the hairpin loops are sets of unpaired bases which connect to only one stem and occur at the end of the some sections. The multi-loop occurs among more than two stems while internal loops exist between only two stems. The internal loop is defined as two strands of unpaired bases occuring between two stems. The bulges are similar to internal loop it occurs between two stems, the only difference is that one strand's bases are unpaired.

The RNA is encoded as a graph using a straightforward representation. Each vertex represents an RNA residue and is labelled with the base code (A, G, C, U or a rare non-standard base). An adjacency matrix encodes the graph edges, which join all residues in sequence and any base pairs. Base-pairs are detected using the X3DNA program 'find-pairs' [22]. The vertices are also labelled with a geometric 'type' label; 1 if it is part of a base pair, 2 if it is unpaired but within 6.5 Å of another base, and 3 otherwise. Two sets of 3D coordinates are also provided, firstly for the backbone position of the residue (the position of atom C3') and secondly the centroid position of the base. From this data, information about the secondary and tertiary structure can be inferred. A set of classification labels from the 8 classes in Table 1 is also provided.

5 Classification Methods

In this section, we present the results of applying some standard classification algorithms on our database. These are sequence alignment (SA), the shortest path kernel (SP), all-paths and cycles kernel (APC), and the Weisfeiler Lehman optimal assignment method (WL-OA). These methods are described briefly in the following sections. SP and APC have an explicit embedding in vector space, which is used. WL-OA produces a kernel matrix which is embedded using kernel embedding. Sequence alignment produces a distance measure which is embedded using multi-dimensional scaling (negative dimensions use the absolute value [23]).

5.1 Sequence-Based Methods

Sequence-based methods mostly use in the study of the DNA. We have applied this method on the primary structure of the RNA for comparison. The sequence of the RNA is the primary structure of the RNA. The nucleotides are represented as strings such as adenine (A), guanine (G), cytosine (C), and uracil (U). The Needleman - Wunsch algorithm [7] to align the strings (A, G, C, U), and Jukes - Cantor score used to calculate the distances between RNA sequences, which are significantly larger than amino-acid sequences. The running time of this algorithm is therefore quite high. The equation of the Jukes - Cantor score:

$$d = -\frac{3}{4}\log(1 - \frac{4}{3}p).\tag{1}$$

Here, p denotes the distance between them in terms of the fraction of letters which differ.

5.2 Weisfeiler-Lehman Optimal Assignment (WL-OA)

The Weisfeiler-Lehman optimal assignment (WL-OA) graph kernel [24] is a state-of-the-art method for labelled graph comparison which utilises an optimal assignment kernel with the labelling generated by the WL method [8]. The method generated vertex labels using the WL label refinement process, with initial labelling corresponding to the RNA vertex labels. An implicit optimal assignment is sought which minimises the labelling difference, and the kernel value is the count of label differences for this optimal assignment. WL-OA performs favourably compared to state-of-art graph kernel in a wide range of datasets.

5.3   Shortest Path Embedding

The walk kernel counts the similar walk in graph pairs. The shortest path kernel (SPK) [25] is a type of walk kernel which counts only the shortest walks between each pair of nodes in a graph:

$$K_{SP}(G,H) = \sum_{p_i \in SP(G)} \sum_{p_j \in SP(H)} K_B(p_i, p_j) \tag{2}$$

where $SP(.)$ is the set of shortest paths in a graph and K_B is a base kernel which compares paths. In the case that the base kernel is the delta kernel, this has an explicit embedding as the histogram of the shortest paths, where each path is denoted by its sequence of labels. This method also called shortest path embedding. Each labelled shortest path in the molecule is generated, and the embedding is a histogram of these paths.

5.4   All Paths and Cycles Embedding(APC)

The all-paths graph kernel is a recently proposed extension to the shortest path kernel, which counts all paths (up to a maximum length).

$$K_{APC}(G,H) = \sum_{p_i \in PC(G)} \sum_{p_j \in PC(H)} K_B(p_i, p_j) \tag{3}$$

Here, $PC(G)$ denotes a set of all paths and simple cycles (a cycle is $v_1 v_2 ... v_n v_1, v_1 < 1 < v_n$ is distinct) on G and $K_B(.,.)$ is a base kernel for paths [26]. In order to evaluate this kernel in a computationally efficient way, the maximum path length and the number of distinct labels must be limited. We therefore label bases with three labels G/C, A/U and other. This is embedded in the same way as the shortest path kernel, as a histogram of distinct labelled paths.

6   Results

In this work, we presented our graph-based RNA dataset. We classified this dataset in 8 type of RNA categories as listed in Table 1. We also demonstrate here that graph based methods can be used to classify RNA molecules. To evaluate the effect of different types of structure (broadly corresponding to the primary, secondary and tertiary structure), we include information from the topology, sequence, and the geometry of shape of the RNA. *Seq* includes the graph edges corresponding to the sequence only, and the base labels. *Top* includes the graph edges (including the sequence and base-pair edges) but no base labels. *Geo* adds additional labels to the bases corresponding to the geometry type labels described earlier. The combinations are the union of these sources of information.

Then, we have applied graph-based embedding methods and a classifier to find the most effective methods on our RNA dataset to determine the accuracy.

Table 2. Classification accuracies for the RNA dataset using Weisfeiler Lehman Optimal Assignment (WL-OA), Shortest Path (SP), All-Paths and Cycle methods (APC), Sequence Alignment (SA), and a variety representations

	Seq only	Top only	Geo only	Seq + Top	Geo + Top	Geo + Seq	All
WL-OA	**92.0**	73.1	**86.8**	**92.4**	**87.1**	89.5	90.2
SP	91.3	79.5	86.7	91.1	86.7	**91.1**	**90.8**
APC	90.3	**85.4**	84.3	89.9	85.5	...	...
SA	89.2	...	...	...	...	...	...

We tried a variety of classifiers; Subspace KNN, Subspace Discriminant, Linear discriminant, Boosted Trees, Cosine KNN and Bagged Trees. In our experiments, Subspace KNN outperform the results with APC methods, and SP methods on three source of information of the RNA. Bagged Trees outperform best results with SA method. On the other hand, WL-OA performed its best results from variety methods on a variety of representations. Subspace KNN performs best result on Geometry Label, Geometry Shape of the RNA + Topology, Geometry + Residue Label, and All Label; Subspace Discriminant and Linear Discriminant demonstrate best results on Residue Label, and Residue + Topology Label; Boosted Trees performs best result on Topology Label (Table 2).

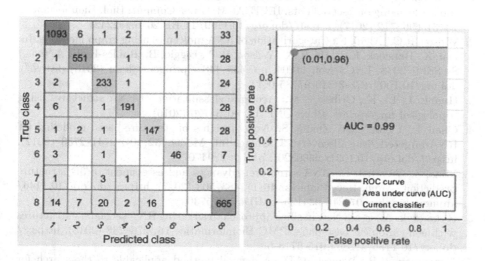

Fig. 2. Confusion Matrix and ROC Curve on WL-OA method on Sequence + Topology Label

The experimental results on the RNA dataset represents that, Weisfeiler-Lehman Optimal Assignment (WL-OA) methods outperform the Shortest Path (SP) method, All Path Cycle (APC) method, and Sequence Alignment (SA) method. The accuracy with WL-OA increased up to 92.4%, which is the best performance in all methods. The results here largely support the results of [2] in that all graph based methods outperform sequence alignment even when only the sequence data is available. It is clear that the base labels are important to the classification, with a drop-off of between 4–10% when they are not included. The all-paths kernel can only be evaluated with three labels, which means that it cannot be used on the experiments with rich label sets and explains the lower performance overall.

7 Conclusion

In this work, we have described graph based methods and kernel based methods for encoding RNA molecules. Then, we applied these methods on our RNA dataset using MATLAB classifiers. We have demonstrated that, graph kernels can be used to classify RNA in high accuracy. We received the best results on the Residue Label (sequence information) and Topology Label using WL-OA with an accuracy of 92.4%. We received the worst results on nucleotide sequences using general Sequence Alignment method with an accuracy of 89.2%.

References

1. Shabash, B., Wiese, K.C.: RNA visualization: relevance and the current state-of-the-art focusing on pseudoknots. IEEE/ACM Trans. Comput. Biol. Bioinformatics **14**(3), 696–712 (2017). https://doi.org/10.1109/TCBB.2016.2522421
2. Wilson, R.C., Algul, E.: Categorization of RNA molecules using graph methods. In: Bai, X., Hancock, E.R., Ho, T.K., Wilson, R.C., Biggio, B., Robles-Kelly, A. (eds.) S+SSPR 2018. LNCS, vol. 11004, pp. 439–448. Springer, Cham (2018). https://doi.org/10.1007/978-3-319-97785-0_42
3. Huang, J., Li, K., Gribskov, M.: Accurate classification of RNA structures using topological fingerprints. PLoS ONE **11**, e0164726 (2016)
4. Chen, L., Calin, G.A., Zhang, S.: Novel insights of structure-based modeling for RNA-targeted drug discovery. J. Chem. Inf. Model. **52**(10), 2741–2753 (2012). https://doi.org/10.1021/ci300320t. pMID: 22947071
5. Miao, Z., Westhof, E.: RNA structure: advances and assessment of 3D structure prediction. Annu. Rev. Biophys. **46**(1), 483–503 (2017). https://doi.org/10.1146/annurev-biophys-070816-034125. pMID: 28375730
6. Rybarczyk, A., et al.: New in silico approach to assessing RNA secondary structures with non-canonical base pairs. BMC Bioinformatics **16**, 276–288 (2015). https://doi.org/10.1186/s12859-015-0718-6
7. Needleman, S.B., Wunsch, C.D.: A general method applicable to the search for similarities in the amino acid sequence of two proteins. J. Mol. Biol. **43**(3), 443–453 (1970)
8. Shervashidze, N., Schweitzer, P., van Leeuwen, E.J., Mehlhorn, K., Borgwardt, K.M.: Weisfeiler-Lehman graph kernels. J. Mach. Learn. Res. **12**, 2539–2561 (2011). http://dl.acm.org/citation.cfm?id=2078187
9. Vert, J.-P.: The optimal assignment kernel is not positive definite (2008). ArXiv e-prints http://adsabs.harvard.edu/abs/2008arXiv0801.4061V

10. Lodhi, H.: Computational biology perspective: kernel methods and deep learning. Wiley Interdisc. Rev. Comput. Stat. **4**(5), 455–465. https://doi.org/10.1002/wics. 1223
11. What is fasta format? https://zhanglab.ccmb.med.umich.edu/FASTA/
12. Shelton, J.M., Brown, S.J.: Fasta-O-Matic: a tool to sanity check and if needed reformat fasta files (2015). bioRxiv https://www.biorxiv.org/content/early/2015/ 08/21/024448
13. Protein data bank contents guide: atomic coordinate entry format description. Wwpdb.org. http://www.wwpdb.org/documentation/file-format-content/ format33/v3.3.html
14. Protein data bank Japan. Pdbj.org. https://pdbj.org
15. Nucleic acid database (NDB). Ndbserver.rutgers.edu. http://ndbserver.rutgers. edu/
16. RCSB PDB. Rcsb.org. https://www.rcsb.org
17. Klosterman, P., Tamura, M., Holbrook, S., Brenner, S.: SCOR: a structural classification of RNA database. Nucleic Acids Res. **30**, 392–394 (2002)
18. Chojnowski, G., Walen, T., Bujnicki, J.M.: RNA bricks - a database of RNA 3D motifs and their interactions. Nucleic Acids Res. **42**(D1), D123–D131 (2014). http://dx.doi.org/10.1093/nar/gkt1084
19. Ray, S.S., Halder, S., Kaypee, S., Bhattacharyya, D.: HD-RNAS: an automated hierarchical database of RNA structures. Front. Genet. **3**, 59 (2012). https://www.frontiersin.org/article/10.3389/fgene.2012.00059
20. York RNA Graph Dataset. https://www.cs.york.ac.uk/cvpr/RNA.html
21. Antczak, M., et al.: RNApdbee 2.0: multifunctional tool for RNA structure annotation. Nucleic. Acids Res. **46**(W1), W30–W35 (2018). https://doi.org/10.1093/ nar/gky314
22. 3DNA: a suite of software programs for the analysis, rebuilding and visualization of 3-dimensional nucleic acid structures. x3dna.org. http://x3dna.org/
23. Duin, R.P.W., Pękalska, E., Harol, A., Lee, W.J., Bunke, H.: On euclidean corrections for non-euclidean dissimilarities. In: da Vitoria, L.N., et al. (eds.) SSPR/SPR 2008. LNCS, vol. 5342, pp. 551–561. Springer, Heidelberg (2008). https://doi.org/ 10.1007/978-3-540-89689-0_59
24. Kriege, N.M., Giscard, P.-L., Wilson, R.C.: On valid optimal assignment kernels and applications to graph classification. In: Advances in Neural Information Processing Systems, pp. 1615–1623 (2016)
25. Borgwardt, K.M., Kriegel, H.: Shortest-path kernels on graphs. In: Proceedings of the 5th IEEE International Conference on Data Mining (ICDM 2005), 27–30 November 2005, Houston, pp. 74–81 (2005). https://doi.org/10.1109/ICDM.2005.132
26. Giscard, P.-L., Wilson, R.C.: The all-paths and cycles graph kernel. arXiv preprint arXiv:1708.01410 (2017)

Event Prediction Based on Unsupervised Graph-Based Rank-Fusion Models

Icaro Cavalcante Dourado[1]([✉])(iD), Salvatore Tabbone[2],
and Ricardo da Silva Torres[1](iD)

[1] Institute of Computing, University of Campinas (UNICAMP), Campinas, Brazil
{idourado,rtorres}@ic.unicamp.br
[2] Université de Lorraine-LORIA UMR 7503, Vandoeuvre-lès-Nancy, France
tabbone@loria.fr

Abstract. This paper introduces an unsupervised graph-based rank aggregation approach for event prediction. The solution is based on the encoding of multiple ranks of a query, defined according to different criteria, into a graph. Later, we embed the generated graph into a feature space, creating fusion vectors. These vectors are then used to train a predictor to determine if an input (even multimodal) object refers to an event or not. Performed experiments in the context of the flooding detection task of the MediaEval 2017 shows that the proposed solution is highly effective for different detection scenarios involving textual, visual, and multimodal features, yielding better detection results than several state-of-the-art methods.

Keywords: Rank aggregation · Graph-based rank fusion ·
Graph embedding · Event prediction

1 Introduction

Nowadays event analysis and especially disaster event is a hot topic that attracts a lot of attention in many parts of the world. Among the different natural disasters, flooding event is one of the most observed and harmful. In this context, natural disaster monitoring is a fundamental task to create prevention strategies, as well as to help authorities to act in the control of damages. Many research works have been proposed in the literature combining heterogeneous data sources (remotely sensed information and social media) to analyze natural disasters. In [26], authors point out the benefit to explore and fuse multi-modal features with different models. Moreover, combining different kinds of features (local vs holistic) improve substantially the retrieval precision [23], and most retrieval fusion approaches are based on rank fusion [7,9,23].

This paper introduces an unsupervised graph-based rank aggregation approach for event prediction. We propose a late-fusion method based on the representation of multiple ranks, defined according to different criteria, into a graph.

© Springer Nature Switzerland AG 2019
D. Conte et al. (Eds.): GbRPR 2019, LNCS 11510, pp. 88–98, 2019.
https://doi.org/10.1007/978-3-030-20081-7_9

Graphs provide an efficient representation of arbitrary structures and inter-relationships among different elements of a model. We embed the generated graph into a feature space, creating fusion vectors. Next, a regressor is trained to predict if an input multimodal object refers to a target event or not, following their fusion vectors.

Experimental results for the flooding detection task of the MediaEval 2017 demonstrate that the proposed solution is robust for different detection scenarios involving textual, visual, and multimodal features, yielding better detection results than state-of-the-art methods. While most previous initiatives for event prediction are solely based on either CNN-based descriptors in isolation, feature concatenation [4,8,10,11,16,21], or graph-based early-fusion [3,22], our method works on top of state-of-the-art descriptors for both visual and textual data, properly designed for the task, along with an extension of a competitive late-fusion approach for generic retrieval tasks [9]. It also has the advantage to be unsupervised and provides a robust way to leverage multiple multimodal rankers.

2   Preliminaries

Let a *sample* s be any digital object, such as a document, an image, a video, or even a hybrid (multimodal) object. A sample is characterized by a *descriptor*, $\mathcal{D}$, which relies on a particular point of view to describe s as a vector, graph or another data structure $\epsilon(s)$. Descriptions allow samples to be compared to each other. Therefore, descriptors are the basis for retrieval and learning models.

A *comparator* $\mathcal{C}$, applied over a tuple $(\epsilon(s_i), \epsilon(s_j))$, produces a *score* $\varsigma \in \mathbb{R}^+$ (e.g., the Euclidean distance or the cosine similarity). Either similarity or dissimilarity functions can be used to implement $\mathcal{C}$. A query sample, or just *query q*, refers to a particular sample taken as an input object in the context of a search, whose purpose is to retrieve *response items* from a *response set* (S) according to relevance criteria. A response set $S = \{s_1, s_2, \ldots, s_n\}$ is a collection of n *samples*, where n is the collection size. A *ranker* is a tuple $R = (\mathcal{D}, \mathcal{C})$, which is employed to compute a rank τ for q, denoted by τ_q to distinguish its corresponding query. A rank is a permutation of $S_L \subseteq S$, where $L \ll n$ in general, such that τ_q provides the most similar – or equivalently the least dissimilar – samples to q from S, in order. L is a cut-off parameter.

A ranker establishes a ranking system, but different descriptors and comparators can compose rankers. Besides, descriptors are commonly complementary, as well as comparators. For m rankers, $\{R_1, R_2, \ldots, R_m\}$, used for query retrieval over a collection S, for every query q we can obtain $\mathcal{T}_q = \{\tau_1, \tau_2, \ldots, \tau_m\}$, from which a *rank aggregation* function f produces a combined rank $\tau_{q,f} = f(\mathcal{T}_q)$, presumably more effective than the individual ranks $\tau_1, \tau_2, \ldots, \tau_m$.

Descriptors have specific pros and cons, because each one focuses on a certain aspect of the data: one may be particularly specialized for either object detection, scene detection, corner detection, keywords, etc. For this reason, descriptors often provide complementary views, when adopted in combination. *Early-fusion* methods emphasize the generation of composite descriptions for samples,

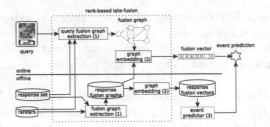

Fig. 1. Proposed graph-based rank fusion for event prediction.

whereas *late-fusion* methods perform a combination of techniques focused on a target problem. Majority voting of classifiers and rank aggregation functions are examples of late-fusion methods. Late-fusion methods are especially useful when the raw data from the objects are not available, and are potentially more effective than early-fusion methods because they are specifically designed or optimized for the problem being solved.

3   Graph-Based Rank Fusion for Event Prediction

Figure 1 presents an overview of our method – an event predictor based on rank-fusion graphs. The solution is composed of three main generic components, briefly described here and detailed in the following sections. Two stages are adopted. The *offline* stage comprehends the modeling of the response set in terms of multiple rankers, as fusion graphs and then as fusion vectors. This step performs a graph embedding of fusion graphs and the training of an event prediction model. The *online* stage refers to the event prediction preceded by a rank-based late-fusion approach. The offline stage is performed only once, while the online stage is performed per prediction. The first two components – *fusion graph extraction* and *graph embedding* – are used in both stages.

The *fusion graph extraction* component (component 1 in the figure) generates a fusion graph $\mathcal{G}$ for a given query sample q. $\mathcal{G}$ consists of an aggregated representation of multiple ranks for q, thus capturing and correlating information of multiple ranks. This formulation is presented in Sect. 3.1. *Graph Embedding* (2) projects fusion graphs into a vector space model, producing a corresponding fusion vector $\mathcal{V}$ for $\mathcal{G}$. We propose an embedding formulation in Sect. 3.2. At the end, an *event predictor* (3) is built based on the response fusion vectors, in order to predict for queries (also modeled as fusion vectors). This component is detailed in Sect. 3.3.

3.1   Fusion Graph Extraction

This component produces a fusion graph $\mathcal{G}$ for a given query sample q, also in terms of m rankers and n response items. A fusion graph is a graph-based encoding of multiple ranks for q, that encapsulates and correlates ranks.

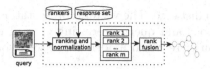

Fig. 2. Extraction of a fusion graph (adapted from [9]).

We follow the fusion graph formulation from [9], referred to as *FG*, that defines a mapping function $q \rightarrow \mathcal{G}$, based on its ranks $\tau \in \mathcal{T}_q$ and ranks' inter-relationships. The proposed formulation also includes a dissimilarity function for *FG*, and a retrieval model based on fusion graphs. Here, however, we focus on the definition of *FG*, extending its use as part of a rank-based late-fusion approach for event prediction, without these components.

The process is illustrated in Fig. 2. Given a query q, m rankers and a response set of size n, m ranks are generated. These ranks are then normalized to allow for producing the fusion graph $\mathcal{G}$ for q. In rank normalization, the scores in the ranks generated by dissimilarity-based comparators are converted to similarity-based scores. Besides, all ranks have their scores rescaled to the same interval. $\mathcal{G}$, for q, includes all response items from each $\tau_q \in \mathcal{T}_q$, as vertices. Vertices are connected by taking into account the degree of relationship between their corresponding response items, and the degree of their relationships to q.

3.2 Fusion Graph Embedding

Let $\mathbb{G} = \{\mathcal{G}_1, \mathcal{G}_2, \ldots, \mathcal{G}_n\}$ be the fusion graph set related to the response set of a certain collection. From $\mathbb{G}$, a fusion graph embedding function $\mathcal{E}$ defines a vector space model in order to project a fusion graph $\mathcal{G}$ into that space as a fusion vector $\mathcal{V}$, i.e. $\mathcal{V} = \mathcal{E}(\mathcal{G})$ for any $\mathcal{G}$.

$\mathcal{E}$ can be defined by unsupervised or supervised approaches. We propose a straightforward yet effective unsupervised approach, and we leave the investigation of supervised formulations for future work. For the following definitions, let $w(v)$ be the weight of the vertex v, if $v \in \mathcal{G}$, otherwise 0. Similarly, let $w(e)$ be the weight of the edge e, if $e \in \mathcal{G}$, otherwise 0. Also, let N be the dimensionality of the vector space model defined by $\mathcal{E}$, such that $\mathcal{V} \in \mathbb{R}^N$.

$\mathcal{E}_V$ is an embedding which derives $\mathcal{V}$ from the vertices of $\mathcal{G}$. There is one vector attribute relative to each response object, therefore $N = |\mathbb{G}|$. $\mathcal{V}$ is derived from $\mathcal{G}$ such that $|\mathcal{V}| = N$, $\mathcal{V}[i]$ is the importance value of the i-th attribute, and $\mathcal{V}[i] = w(v_i)$. Despite the vector space increases linearly to the collection size, the resulting fusion vectors are mainly sparse, i.e., composed of few non-zero entries, which makes this embedding formulation simple and efficient in practice.

A fusion vector is a concise representation of multi-ranked objects, and allows efficient similarity computation and representation storage, being therefore suitable for search and classification tasks. Dissimilarity scores between fusion vectors can be obtained by traditional vector comparators, ranging from correlation metrics to traditional distances and dissimilarity functions, such as Jaccard, cosine, or the Euclidean distance. The vector representation is important to promote efficient comparisons and query processing time of multi-ranked objects, in contrast to prediction or retrieval models based solely on fusion graphs.

3.3 Event Prediction from Fusion Vectors

Fusion vectors allow the creation of predictors (e.g., binary or multi-labeled classifiers), regressors, and also ad-hoc retrieval systems, depending on the underlying demanded task. In this work, we adopt them to model an event predictor, where training objects, associated with ground-truth information, are used to train an estimator for a certain input object be considered an event or not.

Let S be a training corpus of size n. An event predictor can be modeled as a regression model, $Y \approx f(X, \beta)$, where f is an approximation function, X are the independent variables, Y is the dependent variable (target), and β are unknown parameters. A learning model explores S to find a f that minimizes a certain error metric. The training samples are generally labeled, so Y may be categorical. Still, a regressor can be built, as $E(Y|X) = f(X, \beta)$, so that posterior probabilities are inferred in order to estimate a confidence of a sample to refer to an event of not. X, in our case, refers to the fusion vectors, acting as variables that describe the samples in terms of their multiple multimodal ranks.

4 Experimental Evaluation

We present, in this section, the experimental protocol used to evaluate the method, and the results achieved comparatively to state-of-the-art baselines.

4.1 Validation Scenario

We validate our method on the task MediaEval 2017 Disaster Image Retrieval from Social Media [5], which requests event prediction models to infer whether images and/or texts refer to flood events or not. A development set (devset) is provided, consisting of 5,280 sample images and their textual metadata, labeled as either flood (1) or non-flood (0). A test set of 1,320 samples is also provided. The dataset provides one image per sample, along with pre-extracted feature vectors by 9 classical image content descriptors, such as Auto color correlogram (ACC) [13] and Color and Edge Directivity Descriptor (CEDD) [6]. Furthermore, textual data, composed of title, description, and tags, are provided.

Three evaluation scenarios are designed. In the first one, called "visual," only visual data are used. In the second scenario, called "textual," only textual data are used. In the third scenario, called "multimodal," both visual and textual data are expected to be used. The correctness is evaluated, over the test set, by the metric Average Precision at K (AP@K) at various cutoffs (50, 100, 250, 480), and by their mean value (mAP).

Although the task may be seen as a multimodal binary classification problem, the evaluation metrics require ranking-based solutions, or equivalently confidence-level regressors, so that the first positions are the most likely to refer to a flood event. We model our solution as a rank-fusion approach, as for a retrieval system, followed by a flood estimator based on rank-fusion vectors. This approach intends to validate our hypothesis that unsupervised graph-based rank aggregation functions can lead to effective event prediction models.

We selected three visual descriptors and three textual descriptors, for individual analysis in the designed evaluation scenarios, and to evaluate different possibilities of rank-fusion aggregations. The descriptors compose rankers, which are employed to generate ranks in our rank aggregation formulation. Our method varies with respect to which rankers are used to compose late-fusion representation, whether visual rankers, textual ranks, or even combinations of visual and textual rankers for the multimodal scenario are applied.

For the event predictor component, we adopt SVR, an L2-regularized logistic regression based on linear SVM in its dual form, with probabilistic output scores, and trained over the fusion vectors from devset. Probabilistic scores are used so that we can sort the test samples by confidence expectancy of being flood.

Based on the textual metadata, we adopt the following descriptors:

- Bag of Words (BoW) with Term Frequency (TF) weighting;
- 2grams with TF weighting;
- 300-dimensional doc2vec [15] pre-trained on English Wikipedia dataset, of about 35M documents and dumped at 2015-12-01.

The doc2vec model promotes document-level embeddings for texts, and it is based on word embeddings [17], a preliminary work that assigns vector representations for words in order to capture their semantic relationships. We refer to these three adopted descriptors as *BoW*, *2grams*, and *doc2vecWiki*, and we perform the same preprocessing steps for them: lower case conversion, digit and punctuation removal, and English stop word removal. For *BoW* and *2grams*, we also apply Porter stemming.

Despite the visual features provided, we ended up choosing descriptors based on Convolution Neural Networks (CNN). In preliminary analysis, we noticed that CNN-based descriptors surpassed most classical descriptors by large margins. We adopt the following state-of-the-art image descriptors:

- *ResNet50IN*: 2048-dimensional average pooling of the last convolutional layer of ResNet50 [12], pre-trained on ImageNet [19], a dataset of about 14M images labeled for object recognition;
- *VGG16P365*: 512-dimensional average pooling of the last convolutional layer of VGG16 [20], pre-trained on Places365-Standard [25], a dataset of about 10M images of labeled scenes;
- *NASNetIN*: 2048-dimensional average pooling of the last convolutional layer of NASNet [27], pre-trained on ImageNet dataset.

For the deep networks used for visual feature extraction, as well as in the textual feature extraction with doc2vec, we take advantage of pre-trained models. This practice, known as transfer learning, has been effective in many scenarios [14], and it is also particularly beneficial for datasets that are not large enough to generalize the training of such large architectures, as in our case.

Because the problem requires prediction of flood images, we prioritize, in the selection of visual descriptors, datasets for pre-training that focus on images of scenes, aiming at better generality to the target problem.

Table 1. Base results of the chosen descriptors, along with a SVR regressor.

(a) Visual.

Descriptor	AP@50	AP@100	AP@250	AP@480	mAP
ResNet50IN	100.00	98.90	98.02	85.92	95.71
NASNetIN	100.00	100.00	96.01	85.60	95.40
VGG16P365	100.00	97.74	93.65	84.59	94.00

(b) Textual.

Descriptor	AP@50	AP@100	AP@250	AP@480	mAP
BoW	81.85	78.62	72.29	65.51	74.57
2grams	82.01	76.58	73.63	65.40	74.43
doc2vecWiki	77.06	77.40	71.86	64.72	72.76

From the chosen descriptors, rankers are defined as tuples of (descriptor, comparator), where the comparator corresponds to a dissimilarity function. We define the ranker for each descriptor by choosing appropriate comparators. For the textual descriptors BoW and 2grams, we adopt the Weighted Jaccard distance, defined as $1 - J(\mathbf{u}, \mathbf{v})$, where J is the Ruzicka similarity metric (Eq. 1). Jaccard is a well-known and widely-used comparison metric for textual classic descriptors, specially for short texts, as in our case. For the remaining descriptors, we choose the Pearson correlation distance, defined as $1 - \rho(\mathbf{u}, \mathbf{v})$ (Eq. 2), which is a general-purpose metric due to its suitability for highly dimensional data and scale invariance.

$$J(\mathbf{u}, \mathbf{v}) = \frac{\sum_i \min(u_i, v_i)}{\sum_i \max(u_i, v_i)} \tag{1}$$

$$\rho(\mathbf{u}, \mathbf{v}) = \frac{(\mathbf{u} - \bar{u}) \cdot (\mathbf{v} - \bar{v})}{\|(\mathbf{u} - \bar{u})\|_2 \|(\mathbf{v} - \bar{v})\|_2} \tag{2}$$

4.2 Effectiveness Results

We report, in Tables 1a and b, the results in the visual and textual scenarios, achieved by the three visual and textual selected descriptors, along with a SVR regressor. We take these results as an initial baseline.

As the task only mentioned AP@480 and mAP in their leaderboard, we focus our discussions on these two metrics. The correctness for the visual scenario is already high within these baselines, around 85% in AP@480. In the textual scenario, AP@480 is around 65%, which suggests more room for improvement.

Table 2. Flood detection based on **visual** features.

Method	AP@50	AP@100	AP@250	AP@480	mAP
FV-ResNet50IN+NASNetIN+VGG16P365	100.00	100.00	98.55	**88.41**	**96.74**
FV-ResNet50IN+NASNetIN	100.00	100.00	99.00	87.24	96.56
S. Ahmad et al. (2017) [2]				86.81	95.73
Bischke et al. (2017) [4]				86.64	95.71
FV-ResNet50IN+VGG16P365	100.00	100.00	97.89	86.40	96.07
K. Ahmad et al. (2017) [1]				84.94	95.11
Avgerinakis et al. (2017) [3]				78.82	92.27
Dao et al. (2017) [8]				77.62	87.87
Nogueira et al. (2017) [18]	96.20	93.69	87.30	74.67	87.96
Lopez-Fuentes et al. (2017) [16]				67.54	70.16
Hanif et al. (2017) [11]				64.90	80.98
Zhao and Larson (2017) [24]				51.46	64.70
Tkachenko et al. (2017) [21]				50.95	62.75
BoKG (Werneck et al. 2018) [22]	81.11				
BoCG (Werneck et al. 2018) [22]	47.94				
Fu et al. (2017) [10]					19.21

Table 3. Flood detection based on **textual** features.

Method	AP@50	AP@100	AP@250	AP@480	mAP
FV-BoW+2grams+doc2vecWiki	100.00	93.88	84.67	**73.81**	**88.09**
FV-BoW+doc2vecWiki	97.56	93.16	83.47	73.74	86.98
FV-BoW+2grams	92.63	88.19	82.11	71.20	83.54
Tkachenko et al. (2017) [21]				66.78	74.37
Hanif et al. (2017) [11]				65.00	71.79
Zhao and Larson (2017) [24]				63.70	75.74
Bischke et al. (2017) [4]				63.41	77.64
Nogueira et al. (2017) [18]	88.24	84.41	72.61	62.80	77.02
Lopez-Fuentes et al. (2017) [16]				61.58	66.38
Dao et al. (2017) [8]				57.07	57.12
Avgerinakis et al. (2017) [3]				36.15	39.90
K. Ahmad et al. (2017) [1]				25.88	31.45
S. Ahmad et al. (2017) [2]				22.83	18.23
Fu et al. (2017) [10]					12.84

For both visual and textual scenarios, we analyze three variants of our method with respect to the input rankers for late-fusion. For the visual scenario, the combinations are ResNet50IN + NASNetIN, ResNet50IN + VGG16P365, and ResNet50IN + NASNetIN + VGG16P365. For the textual scenario, the combinations are BoW + 2grams, BoW + doc2vecWiki, and BoW + 2grams + doc2vecWiki.

As for the multimodal scenario, we investigate some combinations taking one ranker of each type, two of each, and three of each. Six multimodal combinations

Table 4. Flood detection based on **multimodal** features.

Method	AP@50	AP@100	AP@250	AP@480	mAP
FV-ResNet50IN+VGG16P365+ BoW+doc2vecWiki	100.00	100.00	99.57	**90.96**	**97.63**
FV-ResNet50IN+NASNetIN +VGG16P365+BoW+2grams+doc2vecWiki	100.00	100.00	99.50	90.94	97.61
FV-ResNet50IN+VGG16P365+BoW+2grams	100.00	100.00	99.60	90.68	97.57
FV-ResNet50IN+NASNetIN+BoW+2grams	100.00	100.00	99.13	90.54	97.42
Bischke et al. (2017) [4]				90.45	97.40
FV-ResNet50IN+NASNetIN+BoW+doc2vecWiki	100.00	100.00	99.08	90.00	97.27
FV-ResNet50IN+BoW	100.00	100.00	99.11	89.09	97.05
Nogueira et al. (2017) [18]	100.00	100.00	97.76	85.85	95.90
Dao et al. (2017) [8]				85.41	90.39
S. Ahmad et al. (2017) [2]				83.73	92.55
Lopez-Fuentes et al. (2017) [16]				81.60	83.96
Zhao and Larson (2017) [24]				73.16	85.43
Tkachenko et al. (2017) [21]				72.26	80.87
Avgerinakis et al. (2017) [3]				68.57	83.37
Hanif et al. (2017) [11]				64.60	80.84
K. Ahmad et al. (2017) [1]				54.74	68.12
BoKG (Werneck et al. 2018) [22]	86.90				
BoCG (Werneck et al. 2018) [22]	73.85				
Fu et al. (2017) [10]					18.30

are evaluated: ResNet50IN + BoW, ResNet50IN + NASNetIN + BoW + 2grams, ResNet50IN + NASNetIN + BoW + doc2vecWiki, ResNet50IN + VGG16P365 + BoW + 2grams, ResNet50IN + VGG16P365 + BoW + doc2vecWiki, and ResNet50IN + NASNetIN + VGG16P365 + BoW + 2grams + doc2vecWiki.

We present our results achieved for the three scenarios, using the combinations proposed, along with the results of the 11 teams that participated in the competition. We also show the results achieved by [22] in the visual and multimodal scenarios, which relied on early-fusion techniques.

In the visual scenario, only [2,4] performed better, in terms of AP@480 and mAP, than our preliminary base setup, based on individual descriptors along with the SVR regressor. As for the textual scenario, only [21] in 12 initiatives surpassed BoW + SVR in AP@480, and [4,18,24] in mAP. This indicates that descriptors properly selected to the target problem can overcome more complex models, also requiring less effort.

Our method was superior in the visual scenario to all baselines, for two of three proposed variants of ranker combinations. Compared to the strongest baselines considering this scenario, our method presents gains from around 1 to 2% in AP@480, and 1% in mAP. Compared to the visual base results, from the individual descriptors, 3 to 4% in AP@480, and 1 to 2% in mAP.

In the textual scenario, our gains were even more expressive. It was superior in the textual scenario to all related works, for all three proposed variants of ranker combinations. Compared to the strongest baselines, our method presents gains from 5 to 7% in AP@480, and 6 to 11% in mAP. Compared to the textual base results, from the individual descriptors, 6 to 8% in AP@480, and 14 to 16% in mAP. In the multimodal scenario, considered baselines were more competitive.

Again, however, our method presents gains over them, of 0.5% in AP@480 and 0.23% in mAP.

5 Conclusions

In this paper, we introduced a graph-based rank fusion approach for event prediction. Our solution is based on encoding multiple ranks into a graph representation, which is later embedded into a vectorial representation. Next, a regressor is trained to predict if an input multimodal object refer to a target event or not, given their graph-based fusion representations.

The proposed method extends the fusion graphs, first introduced in [9], to the context of supervised tasks, specifically for event prediction. It also proposes a graph embedding mechanism in order to define the fusions vectors, a new late-fusion vector representation that encodes multiple ranks and their inter-relationships automatically. Performed experiments in the context of the flooding detection task at MediaEval 2017 demonstrates that our solution leads to highly effective results overcoming several state-of-the-art solutions.

Future work will focus on investigating the impact of different graph embedding approaches. We also plan to investigate the use of our solution in other multimodal prediction problems.

Acknowledgments. Authors are grateful to CNPq (grant #307560/2016-3), São Paulo Research Foundation – FAPESP (grants #2014/12236-1, #2015/24494-8, #2016/50250-1, and #2017/20945-0) and the FAPESP-Microsoft Virtual Institute (grants #2013/50155-0, and #2014/50715-9). This study was financed in part by the Coordenação de Aperfeiçoamento de Pessoal de Nível Superior - Brasil (CAPES) - Finance Code 001.

References

1. Ahmad, K., Pogorelov, K., Riegler, M., Conci, N., Pal, H.: CNN and GAN based satellite and social media data fusion for disaster detection. In: MediaEval Workshop, Dublin (2017)
2. Ahmad, S., Ahmad, K., Ahmad, N., Conci, N.: Convolutional neural networks for disaster images retrieval. In: MediaEval Workshop, Dublin (2017)
3. Avgerinakis, K., et al.: Visual and textual analysis of social media and satellite images for flood detection@ multimedia satellite task MediaEval 2017. In: MediaEval Workshop, Dublin (2017)
4. Bischke, B., Bhardwaj, P., Gautam, A., Helber, P., Borth, D., Dengel, A.: Detection of flooding events in social multimedia and satellite imagery using deep neural networks. In: MediaEval Workshop, Dublin (2017)
5. Bischke, B., Helber, P., Schulze, C., Venkat, S., Dengel, A., Borth, D.: The multimedia satellite task at MediaEval 2017: emergence response for flooding events. In: MediaEval Workshop, Dublin (2017)
6. Chatzichristofis, S.A., Boutalis, Y.S.: CEDD: color and edge directivity descriptor: a compact descriptor for image indexing and retrieval. In: Gasteratos, A., Vincze, M., Tsotsos, J.K. (eds.) ICVS 2008. LNCS, vol. 5008, pp. 312–322. Springer, Heidelberg (2008). https://doi.org/10.1007/978-3-540-79547-6_30

7. Dao, I.S., Minh, P.Q.N., Kasem, A.: A context-aware late-fusion approach for disaster image retrieval from social media. In: ICMR 2018. ACM (2018)
8. Dao, M.S., Pham, Q.N.M., Nguyen, D., Tien, D.: A domain-based late-fusion for disaster image retrieval from social media. In: MediaEval Workshop, Dublin (2017)
9. Dourado, I.C., Pedronette, D.C.G., da Silva Torres, R.: Unsupervised graph-based rank aggregation for improved retrieval. Inf. Process. Manage. **56**(4), 1260–1279 (2019). https://doi.org/10.1016/j.ipm.2019.03.008. ISSN 0306-4573
10. Fu, X., Bin, Y., Peng, L., Zhou, J., Yang, Y., Shen, H.T.: BMC@MediaEval 2017 multimedia satellite task via regression random forest. In: MediaEval Workshop, Dublin (2017)
11. Hanif, M., Tahir, M.A., Khan, M., Rafi, M.: Flood detection using social media data and spectral regression based kernel discriminant analysis. In: MediaEval Workshop, Dublin (2017)
12. He, K., Zhang, X., Ren, S., Sun, J.: Deep residual learning for image recognition. In: IEEE Conference on Computer Vision and Pattern Recognition, pp. 770–778 (2016)
13. Huang, J., Kumar, S.R., Mitra, M., Zhu, W.J., Zabih, R.: Image indexing using color correlograms. In: Proceedings of IEEE CVPR 1997, pp. 762–768. IEEE (1997)
14. Kornblith, S., Shlens, J., Le, Q.V.: Do better imagenet models transfer better? arXiv preprint arXiv:1805.08974 (2018)
15. Le, Q., Mikolov, T.: Distributed representations of sentences and documents. In: International Conference on Machine Learning, pp. 1188–1196 (2014)
16. Lopez-Fuentes, L., van de Weijer, J., Bolanos, M., Skinnemoen, H.: Multi-modal deep learning approach for flood detection. In: MediaEval Workshop, Dublin (2017)
17. Mikolov, T., Sutskever, I., Chen, K., Corrado, G.S., Dean, J.: Distributed representations of words and phrases and their compositionality. In: Advances in Neural Information Processing Systems, pp. 3111–3119 (2013)
18. Nogueira, K., et al.: Data-driven flood detection using neural networks. In: MediaEval Workshop, Dublin (2017)
19. Russakovsky, O., et al.: Imagenet large scale visual recognition challenge. IJCV **115**(3), 211–252 (2015)
20. Simonyan, K., Zisserman, A.: Very deep convolutional networks for large-scale image recognition. arXiv preprint arXiv:1409.1556 (2014)
21. Tkachenko, N., Zubiaga, A., Procter, R.: WISC at MediaEval 2017: multimedia satellite task. In: MediaEval Workshop, Dublin (2017)
22. Werneck, R.O., Dourado, I.C., Fadel, S.G., Tabbone, S., Torres, R.S.: Graph-based early-fusion for flood detection. In: 25th IEEE ICIP, pp. 1048–1052. IEEE (2018)
23. Zhang, S., Yang, M., Cour, T., Yu, K., Metaxas, D.N.: Query specific rank fusion for image retrieval. IEEE PAMI **37**(4), 803–815 (2015)
24. Zhao, Z., Larson, M.: Retrieving social flooding images based on multimodal information. In: MediaEval Workshop, Dublin (2017)
25. Zhou, B., Lapedriza, A., Khosla, A., Oliva, A., Torralba, A.: Places: a 10 million image database for scene recognition. IEEE Trans. Pattern Anal. Mach. Intell. **40**(6), 1452–1464 (2018). https://doi.org/10.1109/TPAMI.2017.2723009. ISSN 0162-8828
26. Zhou, W., Li, H., Tian, Q.: Recent advance in content-based image retrieval: a literature survey. CoRR abs/1706.06064 (2017)
27. Zoph, B., Vasudevan, V., Shlens, J., Le, Q.V.: Learning transferable architectures for scalable image recognition. arXiv preprint arXiv:1707.07012 (2017)

Generalized Median Graph via Iterative Alternate Minimizations

Nicolas Boria[1](✉), Sébastien Bougleux[1], Benoit Gaüzère[2], and Luc Brun[1]

[1] Normandie Univ, UNICAEN, ENSICAEN, CNRS, GREYC, Caen, France
{nicolas.boria,bougleux}@unicaen.fr, luc.brun@ensicaen.fr
[2] Normandie Univ, INSA ROUEN Normandie, LITIS, Rouen, France
benoit.gauzere@insa-rouen.fr

Abstract. Computing a graph prototype may constitute a core element for clustering or classification tasks. However, its computation is an NP-Hard problem, even for simple classes of graphs. In this paper, we propose an efficient approach based on block coordinate descent to compute a generalized median graph from a set of graphs. This approach relies on a clear definition of the optimization process and handles labeling on both edges and nodes. This iterative process optimizes the edit operations to perform on a graph alternatively on nodes and edges. Several experiments on different datasets show the efficiency of our approach.

Keywords: Median graph · Graph Edit Distance · Optimization

1 Introduction

In a wide variety of scientific domains, attributed graphs provide a powerful structure to represent, process and analyze data. However, determining fundamental tools such as a distance or an average graph is non trivial. Given a space $\mathbb{G}$ of attributed graphs, *Graph Edit Distance* (GED) is a natural choice for comparing graphs [2,16]. It measures the minimal amount of distortion needed to transform a graph into another by means of edit operations. It can be defined as a minimal-path problem which relies on a cost function acting as a metric in $\mathbb{G}$, and rewritten as a special quadratic assignment problem close to the graph matching problem. Computing Graph Edit Distance is NP-Hard and still cannot be solved in a reasonable time for graphs exceeding a dozen of vertices, even for simple cost functions. Therefore, several strategies have been explored to provide tight upper-bounds in polynomial time [16]. Computing a representative of a set of graphs $\mathcal{G} \subset \mathbb{G}$ is even more difficult. It commonly consists in finding a *generalized median graph*, *i.e.* a graph $\bar{G} \in \mathbb{G}$ that minimizes the *sum of distances* (SOD) to all the graphs in $\mathcal{G}$ [10]:

$$\bar{G} \in \arg \min_{G \in \mathbb{G}} \sum_{G' \in \mathcal{G}} d(G, G') \tag{1}$$

© Springer Nature Switzerland AG 2019
D. Conte et al. (Eds.): GbRPR 2019, LNCS 11510, pp. 99–109, 2019.
https://doi.org/10.1007/978-3-030-20081-7_10

where $d : \mathbb{G} \times \mathbb{G} \to \mathbb{R}_+$ denotes Graph Edit Distance. Exact methods are restricted to labeled graphs with particular cost functions or datasets containing a small total number of vertices [5]. To estimate median graphs in a reasonable computational time, several methods reduce the SOD by a local search around an initial candidate graph, by genetic search [10], greedy search based on partitioning vertices of different graphs [9], greedy adaptive search [13], or linearization and discrete optimization [12]. A different strategy is based on graph embedding [3,6–8,14], usually with distances between graphs as coordinates. A representative is more easily computed within this space. Then a median graph is reconstructed by going back to the original space of graphs. While these approaches are able to tackle the complexity of the previous ones, the link with the definition of a generalized median graph is not trivial and difficult to analyze. Other approaches use the relationship between common-labeling and the median graph to derive bounds on the SOD [15], or extend the concept of representative to correspondences between graphs [11].

In this paper, we propose to estimate a generalized median graph by a block coordinate descent that iterates two minimization steps from an initial candidate (Sect. 3): one for updating the SOD w.r.t. edges and attributes on nodes and on edges, and the other w.r.t. distances. The order of the resulting graph is fixed before the descent process by the order of the initial candidate. This candidate is set to a *set-median*, *i.e.* a graph of $\mathcal{G}$ minimizing the SOD ($\mathbb{G}$ restricted to $\mathcal{G}$ in Eq. 1). While the first step of the descent shares similarities with the update presented in [10], the update rules are not the same, and any algorithm can be used to estimate GED in the second step or for initialization. The first empirical results on two datasets (Sect. 4) show on the one hand that the proposed method systematically reduces the SOD associated with the initial candidate, *i.e.* a set-median, and on the other hand that the accuracy of the approximate GED has more impact on the descent than on the computation of a set-median. The following section introduces the expressions we use to facilitate the derivation of the proposed algorithm.

2   Graph Transformations and Graph Edit Distance

We consider simple undirected attributed graphs. An attributed graph G of order n can be encoded by a triplet $(\varphi, A, \varPhi)$ (Fig. 1). The n-tuple $\varphi = (\varphi_i)_i$ associates an attribute (or feature) φ_i of a space $\mathbb{F}_v$ to each integer $i \in [n] = \{1, \ldots, n\}$ (vertices are represented by the set $[n]$). $A \in \{0,1\}^{n \times n}$ is the vertex-vertex adjacency matrix of G, *i.e.* $a_{i,j} = 1$ if there is an edge (i,j), else $a_{i,j} = 0$. $\varPhi = (\phi_{i,j})_{i,j}$ associates an attribute $\phi_{i,j}$ of a space $\mathbb{F}_e$ to each pair $(i,j) \in [n] \times [n]$. When (i,j) is not an edge, $\phi_{i,j}$ can be equal to any value, it does not affect the following expressions. Obviously, A and $\varPhi$ are symmetric. Let $\mathbb{G}$ be the space of all attributed graphs for $\mathbb{F}_v$ and $\mathbb{F}_e$ fixed. In this paper, each space of attributes is restricted to a finite set of positive integer labels, or to the Euclidean space.

A graph $G = (\varphi, A, \varPhi)$ of order n can be transformed into a graph $G' = (\varphi', A', \varPhi')$ of order n' by applying a composition of elementary transformations,

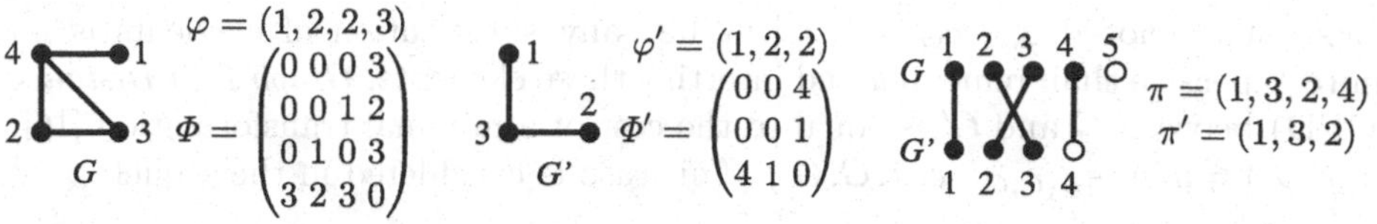

$$\varphi = (1,2,2,3)$$

$$\Phi = \begin{pmatrix} 0 & 0 & 0 & 3 \\ 0 & 0 & 1 & 2 \\ 0 & 1 & 0 & 3 \\ 3 & 2 & 3 & 0 \end{pmatrix}$$

$$\varphi' = (1,2,2)$$

$$\Phi' = \begin{pmatrix} 0 & 0 & 4 \\ 0 & 0 & 1 \\ 4 & 1 & 0 \end{pmatrix}$$

$$\pi = (1,3,2,4)$$
$$\pi' = (1,3,2)$$

Fig. 1. Labeled graphs (label 0 if no edge) and a transformation (π, π') of their vertices. Induced operations on edges: $\phi_{2,3} = \phi_{3,2}$ substituted by $\phi'_{3,2} = \phi'_{2,3}$, $(1,4), (2,4), (3,4)$ removed from G, $(1,2)$ inserted in G from $(1,3)$ in G' with $\phi_{1,2} = \phi_{2,1} = \phi'_{1,3}$.

a.k.a. *edit operations*, to G. An edit operation transforms a graph into another by either removing an element (a vertex or an edge), substituting an attribute attached to an element by another attribute, or by inserting an element and its attribute (between two existing vertices for edges). Moreover, if each element of both graphs is assumed to be involved in exactly one edit operation, the number of operations is minimized, and the transformation of G into G' is fully described by the transformation of the vertices of G into the ones of G'. Here, this transformation, a.k.a. *error-correcting matching* [2,16], is defined as a pair $(\pi, \pi') \in [n'+1]^n \times [n+1]^{n'}$ so that $\pi_i = k \in [n'] \Leftrightarrow \pi'_k = i \in [n]$ (Fig. 1). Each vertex i of G is either substituted by a vertex k of G' ($\pi_i = k$ and $\pi'_k = i$), or removed ($\pi_i = n'+1$). Each vertex k of G' that is not substituted to a vertex of G is inserted ($\pi'_k = n+1$). The transformation of the edges of G into the ones of G' is induced by the transformation of the vertices. The set $\{(i,j) \in [n] \times [n] \mid a_{i,j} = 1 \wedge \pi_i \in [n'] \wedge \pi_j \in [n'] \wedge a_{\pi_i, \pi_j} = 1\}$ defines the substituted edges, the set $\{(i,j) \in [n] \times [n] \mid a_{i,j} = 1 \wedge ((\pi_i \in [n'] \wedge \pi_j \in [n'] \wedge a_{\pi_i, \pi_j} = 0) \vee \pi_i = n'+1 \vee \pi_j = n'+1)\}$ defines the removed edges, and the set $\{(k,l) \in [n'] \times [n'] \mid a'_{k,l} = 1 \wedge ((\pi'_k \in [n] \wedge \pi'_l \in [n] \wedge a_{\pi'_k, \pi'_l} = 0) \vee \pi'_k = n+1 \vee \pi'_l = n+1)\}$ defines the inserted edges. Since π' can be obtained from π, we omit π' for simplicity, and we denote by $\Pi(G, G')$ all the transformations of G to G'.

A transformation $\pi^* \in \Pi(G, G')$ is said to be minimal if its cost is minimal, i.e. if $c(\pi^*, G, G') = \min_{\pi \in \Pi(G,G')} c(\pi, G, G')$, with $c(\pi, G, G') = c_v(\pi, \varphi, \varphi') + \frac{1}{2} c_e(\pi, A, \Phi, A', \Phi')$ the cost for transforming G into G' using π, and

$$c_v(\pi, \varphi, \varphi') = \sum_{i=1}^{n} \delta_{\pi_i} c_{\text{vfs}}(\varphi_i, \varphi'_{\pi_i}) + (1 - \delta_{\pi_i}) c_{\text{vr}} + \sum_{k=1}^{n'} (1 - \delta_{\pi'_k}) c_{\text{vi}} \qquad (2)$$

$$c_e(\pi, A, \Phi, A', \Phi') = \sum_{i=1}^{n} \sum_{j=1}^{n} \delta_{\pi_i \pi_j} a_{i,j} a'_{\pi_i, \pi_j} c_{\text{efs}}\left(\phi_{i,j}, \phi'_{\pi_i \pi_j}\right)$$
$$+ c_{\text{er}} \sum_{i=1}^{n} \sum_{j=1}^{n} \delta_{\pi_i \pi_j} a_{i,j} (1 - a'_{\pi_i, \pi_j}) + (1 - \delta_{\pi_i \pi_j}) a_{i,j}$$
$$+ c_{\text{ei}} \sum_{i=1}^{n} \sum_{j=1}^{n} \delta_{\pi_i \pi_j} (1 - a_{i,j}) a'_{\pi_i \pi_j} + c_{\text{ei}} \sum_{k=1}^{n'} \sum_{l=1}^{n'} (1 - \delta_{\pi'_k \pi'_l}) a'_{k,l}$$
$$(3)$$

the costs for transforming attributed vertices and edges, respectively. $\delta_{\pi_i} = 1$ if $\pi_i \in [n']$, else 0, and $\delta_{\pi_i \pi_j} = \delta_{\pi_i} \delta_{\pi_j}$. Functions $c_{\text{vfs}} : \mathbb{F}_v \times \mathbb{F}_v \to [0, +\infty)$ and $c_{\text{efs}} : \mathbb{F}_e \times \mathbb{F}_e \to [0, +\infty)$ measure costs to substitute vertices and edges. In this paper, the costs for removing and inserting elements are restricted to positive constants, denoted c_{vr}, c_{vi}, c_{er}, c_{ei}. When any substitution of elements is no more expensive than removing and inserting these elements, *Graph Edit Distance* (GED) between G and G' is equal to the cost of a minimal transformation [16]: $d(G, G') = \min_{\pi \in \Pi(G,G')} c(\pi, G, G')$. This case is considered in the sequel.

3 Estimating a Generalized Median Graph

Given a set of graphs $\mathcal{G} = \{G_p\}_p \subset \mathbb{G}$, with $G_p = (\varphi_p, A_p, \phi_p)$ of order n_p, a *generalized median graph* $\bar{G} = (\bar{\varphi}, \bar{A}, \bar{\phi}) \in \mathbb{G}$ of $\mathcal{G}$ minimizes the *sum of distances* (SOD) to the graphs of $\mathcal{G}$ [5,10]: $s(\bar{G}, \mathcal{G}) = \min_{G \in \mathbb{G}} s(G, \mathcal{G})$, with $s(G, \mathcal{G}) = \sum_{G_p \in \mathcal{G}} d(G, G_p) = \sum_{p=1}^{|\mathcal{G}|} \min_{\pi_p \in \Pi(G,G_p)} c(\pi_p, G, G_p)$. We propose to use a block coordinate descent to estimate both $\bar{G}$ and the minimal transformations $(\pi_p)_p$.

3.1 Proposed Algorithm

First, $\bar{G}$ is initialized to a set-median of $\mathcal{G}$, *i.e.* $\bar{G} = \arg \min_{G_p \in \mathcal{G}} s(G_p, \mathcal{G})$. It can be computed in $O(a|\mathcal{G}|^2)$ time [5], where a is the complexity of the algorithm used for computing or estimating GED. This also provides the minimal transformations $(\bar{\pi}_p)_p$ from $\bar{G}$ to the graphs of $\mathcal{G}$. The order $\bar{n}$ of $\bar{G}$ is then fixed, *i.e.* considered as a constant in the optimization process. Then, $(\bar{\varphi}, \bar{A}, \bar{\Phi})$ and $(\bar{\pi}_p)_p$ are alternatively updated as follows:

$$\bar{G} = (\bar{\varphi}, \bar{A}, \bar{\Phi}) \leftarrow \arg \min_{\substack{\varphi \in \mathbb{F}_v^{\bar{n}} \\ A \in \{0,1\}^{\bar{n} \times \bar{n}} \\ \Phi \in \mathbb{F}_e^{\bar{n} \times \bar{n}}}} \sum_{p=1}^{|\mathcal{G}|} c_v(\bar{\pi}_p, \varphi, \varphi_p) + \tfrac{1}{2} c_e(\bar{\pi}_p, A, \Phi, A_p, \Phi_p) \quad (4)$$

$$\forall p \in \{1, \dots, |\mathcal{G}|\}, \quad \bar{\pi}_p \leftarrow \arg \min_{\pi_p \in \Pi(\bar{G}, G_p)} c(\pi_p, \bar{G}, G_p) \quad (5)$$

until convergence, that is, until a stability is reached both in $\bar{G}$ and $(\bar{\pi}_p)_p$. The resolution of the minimization of the sum of distances when the transformations are fixed (Eq. 4) mainly depends on the nature of $\mathbb{F}_v$ and $\mathbb{F}_e$, as well as the form of the cost functions c_{vfs} and c_{vef}. This is detailed later in this section, in particular it can be solved in $O(\bar{n}^2|\mathcal{G}|)$ time under some conditions. The update of the transformations (Eq. 5) consists in solving $|\mathcal{G}|$ times GED problem, so in $O(a|\mathcal{G}|)$ time. Since the order $\bar{n}$ is fixed, and GED can usually be only estimated, the algorithm may not converge to the true generalized median graph.

We assume that an algorithm for computing GED is given, and we focus on the minimization of the sum of distances w.r.t. the graph (Eq. 4). It can be decomposed into two independent minimizations as long as the attributes φ_p and Φ_p are independent for each p, that we consider in this paper:

$$\bar{\varphi} \leftarrow \arg \min_{\varphi \in \mathbb{F}_v^{\bar{n}}} s_v(\varphi), \quad (\bar{A}, \bar{\Phi}) \leftarrow \arg \min_{\substack{A \in \{0,1\}^{\bar{n} \times \bar{n}} \\ \Phi \in \mathbb{F}_e^{\bar{n} \times \bar{n}}}} s_e(A, \phi) \quad (6)$$

with $s_v(\varphi) = \sum_{p=1}^{|\mathcal{G}|} c_v(\bar{\pi}_p, \varphi, \varphi_p)$ and $s_e(\phi, A) = \sum_{p=1}^{|\mathcal{G}|} c_e(\bar{\pi}_p, A, \phi, A_p, \phi_p)$. The minimization of each term is detailed in the two following sections. Note that some results are already presented in [10], in particular for vertices. There are obtained in a different way, allowing to take into account more easily different spaces of attributes and cost functions associated to edit operations.

3.2   Updating Vertex Attributes

Only the cost function c_{vfs} depends on vertex attributes in the expression of c_v (Eq. 2). So the attributes $\bar{\varphi}$ in Eq. 6 are updated by solving the equivalent problem $\arg\min_{\varphi \in \mathbb{F}_v^{\bar{n}}} \sum_{i=1}^{\bar{n}} f_i(\varphi_i)$, with the function $f_i : \mathbb{F}_v \to \mathbb{R}_+$ defined by $f_i(\varphi_i) = \sum_{p=1}^{|\mathcal{G}|} \delta_{\pi_i^p} c_{\text{vfs}}(\varphi_i, \varphi_{\pi_i^p}^p)$. The objective function is a sum of positive and independent terms f_i, so the attributes are updated by:

$$\forall i = 1, \ldots, \bar{n}, \quad \bar{\varphi}_i \leftarrow \arg\min_{\varphi_i \in \mathbb{F}_v} f_i(\varphi_i) \tag{7}$$

The solution depends on $\mathbb{F}_v$ and on the cost function c_{vfs}.

When attributes are labels ($\mathbb{F}_v \subset \mathbb{N}$), the cost for substituting a label $x \in \mathbb{F}_v$ by a label $y \in \mathbb{F}_v$ is defined as $c_{\text{vfs}}(x, y) = c_{\text{vs}}(1 - \delta_{x,y})$, with $c_{\text{vs}} > 0$ a constant, $i.e.$ 0 if the labels are the same, and c_{vs} otherwise. Then f_i can be rewritten as $f_i(\varphi_i) = \sum_{p=1}^{|\mathcal{G}|} \delta_{\pi_i^p} c_{\text{vs}}(1 - \delta_{\varphi_i, \varphi_{\pi_i^p}^p}) = c_{\text{vs}}(|S_i| - h_i^0(\varphi_i))$, where $S_i = \{\pi_i^p \mid \pi_i^p \in [n_p], p = 1, \ldots, |\mathcal{G}|\}$ is the set of vertices that are substituted to i by the mappings π_p, and $h_i^0 : \mathbb{F}_v \to \{0, \ldots, |\mathcal{G}|\} \subset \mathbb{N}$, $h_i^0(\varphi_i) = \sum_{p=1}^{|\mathcal{G}|} \delta_{\pi_i^p} \delta_{\varphi_i, \varphi_{\pi_i^p}^p}$, counts the number of times i is substituted by a vertex having the same label (with zero cost). So the attributes (Eq. 7) are updated by:

$$\forall i = 1, \ldots, \bar{n}, \quad \bar{\varphi}_i \leftarrow \arg\max_{\varphi_i \in \mathbb{F}_v} h_i^0(\varphi_i) \tag{8}$$

Notice that h_i^0 can be pre-computed in $O(|\mathcal{G}|)$ time for each label of $\mathbb{F}_v$. The labels are thus updated for all the vertices of $\bar{G}$ in $O(\bar{n}|\mathcal{G}|)$ time at each iteration.

When $\mathbb{F}_v = \mathbb{R}^m$ is equipped with the scalar product $x^T y = \sum_{k=1}^m x_k y_k$ and the l_2-norm $\|x\| = \sqrt{x^T x}$, the cost for substituting an attribute x by an attribute y is defined by $c_{\text{vfs}}(x, y) = \|x - y\|^2$. In this case, we have: $f_i(\varphi_i) = \sum_{p=1}^{|\mathcal{G}|} \delta_{\pi_i^p} \|\varphi_i - \varphi_{\pi_i^p}^p\|^2$. Any attribute $\bar{\varphi}_i$ satisfies $\nabla f_i(\bar{\varphi}_i) = 0$, $i.e.$ $2\sum_p \delta_{\pi_i^p}(\bar{\varphi}_i - \varphi_{\pi_i^p}^p) = 0$, or:

$$\forall i = 1, \ldots, \bar{n}, \quad \bar{\varphi}_i \leftarrow \frac{1}{\sum_{p=1}^{|\mathcal{G}|} \delta_{\pi_i^p}} \sum_{p=1}^{|\mathcal{G}|} \delta_{\pi_i^p} \varphi_{\pi_i^p}^p = \frac{1}{|S_i|} \sum_{p \in S_i} \varphi_{\pi^p(i)}^p \tag{9}$$

In other words, the optimal attribute for a vertex i is given by the mean attribute of the vertices substituted to i (the set S_i defined in the previous paragraph). Once more, updating all the attributes is done in $O(\bar{n}|\mathcal{G}|)$ time at each iteration.

3.3 Updating Edges and Their Attributes

The edges of $\bar{G}$, and their attributes, are computed at each step of the descent (Eq. 4) by minimizing s_e (Eq. 6). By removing the constant terms in s_e, *i.e.* in c_e (Eq. 3), it is easy to show that the minimization of s_e can be rewritten as:

$$\underset{\substack{A\in\{0,1\}^{\bar{n}\times\bar{n}} \\ \Phi\in\mathbb{F}_e^{\bar{n}\times\bar{n}}}}{\arg\min}\, s_e(A,\phi) = \underset{\substack{A\in\{0,1\}^{\bar{n}\times\bar{n}} \\ \Phi\in\mathbb{F}_e^{\bar{n}\times\bar{n}}}}{\arg\min}\, \sum_{i=1}^{\bar{n}}\sum_{j=1}^{\bar{n}} f_{i,j}(a_{i,j},\phi_{i,j}) \tag{10}$$

with the function $f_{i,j} : \{0,1\} \times \mathbb{F}_e \to \mathbb{R}_+$ defined by:

$$
\begin{aligned}
f_{i,j}(a_{i,j},\phi_{i,j}) &= a_{i,j}\sum_{p=1}^{|\mathcal{G}|}\delta_{\pi_i^p\pi_j^p}\, a_{\pi_i^p,\pi_j^p}^p\, c_{\mathrm{efs}}(\phi_{i,j},\phi_{\pi_i^p,\pi_j^p}^p)\\
&\quad + c_{\mathrm{er}}a_{i,j}\sum_{p=1}^{|\mathcal{G}|} 1 - \delta_{\pi_i^p\pi_j^p}\, a_{\pi_i^p,\pi_j^p}^p\\
&\quad + c_{\mathrm{ei}}(1-a_{i,j})\sum_{p=1}^{|\mathcal{G}|}\delta_{\pi_i^p\pi_j^p}a_{\pi_i^p,\pi_j^p}^p\\
&= a_{i,j}\sum_{p=1}^{|\mathcal{G}|}\delta_{\pi_i^p\pi_j^p}\, a_{\pi_i^p,\pi_j^p}^p\, c_{\mathrm{efs}}(\phi_{i,j},\phi_{\pi_i^p,\pi_j^p}^p)\\
&\quad + c_{\mathrm{er}}a_{i,j}\,(|\mathcal{G}|-|S_{i,j}|) + c_{\mathrm{ei}}(1-a_{i,j})|S_{i,j}|
\end{aligned}
\tag{11}
$$

where $S_{i,j} = \{(\pi_i^p,\pi_j^p)\,|\,\pi_i^p\in[n_p]\wedge\pi_j^p\in[n_p]\wedge a_{\pi_i^p,\pi_j^p}^p = 1,\, p = 1,\dots,|\mathcal{G}|\}$ is the set of edges that are substituted to (i,j) by the mappings π_p. The terms $f_{i,j}$ are positive and independent from each others, so Eq. 10 is equivalent to:

$$\forall(i,j)\in[\bar{n}]\times[\bar{n}],\, i\neq j,\quad (\bar{a}_{i,j},\bar{\phi}_{i,j}) \leftarrow \underset{\substack{a_{i,j}\in\{0,1\} \\ \phi_{i,j}\in\mathbb{F}_e}}{\arg\min}\, f_{i,j}(a_{i,j},\phi_{i,j}) \tag{12}$$

Since $a_{i,j}$ can only take two values, if $a_{i,j} = 0$ (no edge) then $f_{i,j}(0,\phi_{i,j}) = c_{\mathrm{ei}}|S_{i,j}|$ for any $\phi_{i,j}\in\mathbb{F}_e$, and if $a_{i,j} = 1$ then $f_{i,j}(1,\phi_{i,j})$ is minimized for any

$$\phi_{i,j}^* \in \underset{\phi_{i,j}\in\mathbb{F}_e}{\arg\min}\sum_{p=1}^{|\mathcal{G}|}\delta_{\pi_i^p\pi_j^p}\, a_{\pi_i^p,\pi_j^p}^p\, c_{\mathrm{efs}}(\phi_{i,j},\phi_{\pi_i^p,\pi_j^p}^p) \tag{13}$$

By consequence $f_{i,j}$ is minimized for $\bar{\phi}_{i,j} = \phi_{i,j}^*$ and

$$\bar{a}_{i,j} = \begin{cases} 1 & \text{if } f_{i,j}(1,\bar{\phi}_{ij}) < c_{\mathrm{ei}}|S_{i,j}| \\ 0 & \text{else} \end{cases} \tag{14}$$

Solutions are finally obtained by solving Eq. 13. It depends on $\mathbb{F}_e$ and c_{efs}.

When $\mathbb{F}_v\subset\mathbb{N}$ and $c_{\mathrm{efs}}(x,y) = c_{\mathrm{es}}(1-\delta_{x,y})$, with $c_{\mathrm{es}} > 0$ a constant, is the classical cost for labels, then $f_{i,j}$ (Eq. 11) becomes

$$f_{i,j}(a_{i,j},\phi_{i,j}) = a_{i,j}\left(c_{\mathrm{es}}\left(|S_{i,j}|-h_{i,j}^0(\phi_{i,j})\right)+c_{\mathrm{er}}\left(|\mathcal{G}|-|S_{i,j}|\right)\right)+(1-a_{i,j})c_{\mathrm{ei}}|S_{i,j}|$$

where $h_{i,j}^0(x) = \sum_{p=1}^{|\mathcal{G}|}\delta_{\pi_i^p\pi_j^p}a_{\pi_i^p,\pi_j^p}^p\delta_{x,\phi_{\pi_i^p,\pi_j^p}^p}$ counts the number of times (i,j) is substituted by an edge having the label x. Then $\bar{\Phi}$ and $\bar{A}$ are updated for all $(i,j)\in[\bar{n}]\times[\bar{n}]$ by:

$$\bar{\phi}_{i,j} \leftarrow \underset{x\in\mathbb{F}_e}{\arg\max}\, h_{i,j}^0(x) \tag{15}$$

and

$$\bar{a}_{i,j} \leftarrow \begin{cases} 1 & \text{if } h_{i,j}^0(\bar{\phi}_{i,j}) > |\mathcal{G}|\frac{c_{er}}{c_{es}} + |S_{i,j}|\left(1 - \frac{c_{er}+c_{ei}}{c_{es}}\right) \\ 0 & \text{else} \end{cases} \tag{16}$$

Each edge (i,j) is thus labeled with one of the most present labels among the ones substituted to (i,j). Notice that $h_{i,j}^0 : \mathbb{F}_e \to \{0, \dots, |\mathcal{G}|\}$ and $|S_{i,j}|$ can be computed in $O(|\mathcal{G}|)$ time. So $\bar{\Phi}$ and $\bar{A}$ are computed in $O(\bar{n}^2|\mathcal{G}|)$ time.

Unlabeled graphs can be considered as labeled with a unique label, $e.g.$ $\mathbb{F}_e = \{1\}$. In this case $c_{efs} = 0$ and $h_{i,j}^0 = |S_{i,j}|$, so from Eq. 16 $\bar{A}$ can be computed in $O(\bar{n}^2|\mathcal{G}|)$ time by:

$$\bar{a}_{i,j} \leftarrow \begin{cases} 1 & \text{if } |S_{i,j}| > |\mathcal{G}|\frac{c_{er}}{c_{er}+c_{ei}} \\ 0 & \text{else} \end{cases} \tag{17}$$

Remark. Similar results can be derived for directed graphs, other spaces of attributes and other cost functions, for both vertices and edges. Due to limited space, it is restricted here to the cases considered in the experiments.

4   Experimental Results

In order to evaluate the validity of our method, the algorithm was implemented in C++ and tested on the datasets Letter (HIGH) [16] and Monoterpenoides[1], a chemical dataset, on a computer using an intel(R) i7-8700 CPU with 12 parallel threads. The Monoterpenoides dataset has 286 graphs unevenly divided in 8 classes of at least 10 graphs. Both nodes and egdes are labeled, and the average order is 11.003. Edit costs were set to $c_{vs} = c_{es} = 1$ and $c_{vi} = c_{ei} = c_{vr} = c_{er} = 3$.

Remember that, in a first phase, the proposed algorithm (Sect. 3.1) identifies a set-median by computing all pairwise distances in the dataset. These distances are computed through two heuristics: bipartite [16], and IPFP [1]. In a second phase, the algorithm iterates the update of a triplet $(\bar{\varphi}, \bar{A}, \bar{\Phi})$ according to Eq. 6 (*i.e.* for vertices either Eq. 8 for Monoterpenoides or Eq. 9 for Letter, and for edges, Eqs. 15–16 for Monoterpenoides or Eq. 17 for Letter), and the update of the transformations $\bar{\pi}_p$ using either bipartite or IPFP. We denote by mBipartite and mIPFP the multistart counterparts of Bipartite, and IPFP [4], where the number of randomly generated initializations was set to 40.

Table 1 sums up our results regarding SOD. In Letter and Monoterpenoides, respectively 50 and 10 graphs were picked randomly in each class, and each experiment was repeated 50 times. The results presented in Table 1 represent the averages over all classes and all experiments. The four columns SOD SM, t(SM), SOD GM and t(GM) list the SODs and computation times in seconds for the set-median (SM), and the generalized median (GM). Note that t(GM) refers to the computation time of the second phase only. Using state of the art GED heuristics and making the most of the computed transformations $\bar{\pi}_p$ to efficiently perform the descent (conversely to many other approaches which use GED only to evaluate candidate medians, without using the detailed transformations), our

---

[1] GREYC Chemistry dataset: https://brunl01.users.greyc.fr/CHEMISTRY/.

algorithm produces median graphs with SODs much lower than the set-medians' with a very low running time. It is noteworthy that the time dedicated to identify the set-median (first phase) is systematically higher than the one dedicated to the generalized median (second phase). Indeed, $|\mathcal{G}|^2$ distances must be computed in the first phase, while $p|\mathcal{G}|$ distances are computed in the second phase, where p denotes the number of iterations before convergence. In practice, we verified that, in most cases, $p < 2$ on the letter dataset, and $p < 7$ on Monoterpenoides. Interestingly enough, in the hybrid versions of the algorithm (using `Bipartite` in the first phase and `IPFP` in the second phase), the alternate descent still produces median graph with reasonably low SOD while starting from a set-median of lesser quality (*i.e.* with higher SODs).

Table 1. SOD computed using different GED approximations.

Algorithms		Letter (HIGH)				Monoterpenoides			
1st phase	2nd phase	SOD SM	t(SM)	SOD GM	t(GM)	SOD SM	t(SM)	SOD GM	t(GM)
Bipartite	Bipartite	142.69	0.01	87.80	$6*10^{-4}$	402.50	0.002	253.11	$8*10^{-4}$
Bipartite	IPFP	142.87	0.013	87.61	0.003	398.01	0.002	128.45	0.179
IPFP	IPFP	135.99	0.057	87.22	0.003	202.75	0.162	104.11	0.136
mBipartite	mBipartite	142.04	0.014	89.47	$9*10^{-4}$	283.94	0.027	186.15	0.01
mBipartite	mIPFP	142.19	0.018	87.66	0.013	281.14	0.031	83.11	0.545
mIPFP	mIPFP	135.99	0.274	87.23	0.015	106.10	1.159	75.08	0.288

Finally, note that the range between best and worst computed SODs is particularly low on the Letter dataset, while it is rather high on the Monoterpenoides dataset. This seems to indicate that approximate computed distances are close to the optimum in Letter, and far from it in Monoterpenoides.

Picking random trainsets in each class 10% and 30% the size of the class, set-medians and generalized medians were computed for each class, and the classification accuracy of a 1-nn algorithm [5] was evaluated using as training examples: (SM) only the set-median, (GM) only the generalized medians and finally (TS) the whole trainset. Each experiment was repeated 50 times, and Table 2 presents our results, giving the average preprocessing time pt (*i.e.* the time spent in computation of set-medians and generalized medians), as well as classification precisions (denoted by %) and times for all three training examples considered. Note that the GED heuristic used in the second phase of the algorithms were also used in computing distances by the classifier.

Let us note that our approach competes with a 1-nn classification over the whole trainset, especially when all the distances are computed with a more precise heuristic, such as mIPFP. Whenever a precise heuristic is used to compute it, the generalized median appears as a better representative of the class than the set-median. Obviously, classification times are much faster using only the median graphs as training example.

In few cases, the classification accuracy enabled by set-medians is higher than that enabled by generalized medians. This only happens in cases where computed distances and edit-paths are looser approximations, *i.e.* this always happens on the Monoterpenoides dataset with the $m\texttt{Bipartite}$ heuristic used in the initialization phase.

Table 2. Classification results for Letter (HIGH) and Monoterpenoides datasets

Letter (HIGH) dataset

TS	1st phase	2nd phase	pt	% SM	t(SM)	% GM	t(GM)	% TS	t(TS)
10%	$m\texttt{Bipartite}$	$m\texttt{Bipartite}$	0.023	76.42	0.325	82.82	0.325	83.01	5.275
	$m\texttt{Bipartite}$	$m\texttt{IPFP}$	0.195	77.40	5.857	84.16	5.771	83.30	110.48
	$m\texttt{IPFP}$	$m\texttt{IPFP}$	0.447	78.24	5.951	84.60	5.801	82.95	111.84
30%	$m\texttt{Bipartite}$	$m\texttt{Bipartite}$	0.181	79.94	0.251	84.24	0.250	87.24	11.44
	$m\texttt{Bipartite}$	$m\texttt{IPFP}$	0.878	81.83	4.323	86.06	4.234	86.86	239.14
	$m\texttt{IPFP}$	$m\texttt{IPFP}$	3.437	81.59	4.316	86.08	4.245	86.86	240.96

Monoterpenoides dataset

TS	1st phase	2nd phase	pt	% SM	t(SM)	% GM	t(GM)	% TS	t(TS)
10%	$m\texttt{Bipartite}$	$m\texttt{Bipartite}$	0.054	32	0.984	29.44	0.957	51.86	3.830
	$m\texttt{Bipartite}$	$m\texttt{IPFP}$	1.586	53.38	47.96	57.49	51.03	60.69	186.85
	$m\texttt{IPFP}$	$m\texttt{IPFP}$	2.044	54.06	47.31	62.38	48.01	60.69	187.83
30%	$m\texttt{Bipartite}$	$m\texttt{Bipartite}$	0.373	36.39	0.747	34.28	0.732	67.92	8.571
	$m\texttt{Bipartite}$	$m\texttt{IPFP}$	5.148	54.06	36.54	67.79	37.07	75.82	419.81
	$m\texttt{IPFP}$	$m\texttt{IPFP}$	15.38	58.37	36.15	74.12	36.57	75.94	419.31

5   Conclusion

We proposed an innovative general method to compute the generalized median graph based on an alternate gradient descent. We showed its efficiency through experiments on two datasets using different edit-cost structures. Computed graphs have much lower SODs than set-medians, and can efficiently be used as representatives in a clustering framework. Quality of computed graph median increases when using accurate rather than fast GED approximation algorithms as sub-routines, especially in the alternate descent phase, while the initialization phase may use different GED heuristics to reach different time/quality compromises. Future developments regarding this promising method include the extension to new edit-cost structures, as well as the possibility to modify the order of the median graph during the optimization process.

Acknowledgments. This work is supported by Région Normandie through RIN AGAC project.

References

1. Bougleux, S., Gaüzère, B., Brun, L.: Graph edit distance as a quadratic program. In: International Conference on Pattern Recognition, pp. 1701–1706 (2016). https://doi.org/10.1109/ICPR.2016.7899881
2. Bunke, H., Allermann, G.: Inexact graph matching for structural pattern recognition. Pattern Recogn. Lett. 1(4), 245–253 (1983). https://doi.org/10.1016/0167-8655(83)90033-8
3. Chaieb, R., Kalti, K., Luqman, M.M., Coustaty, M., Ogier, J.M., Amara, N.E.B.: Fuzzy generalized median graphs computation: application to content-based document retrieval. Pattern Recogn. 72, 266–284 (2017). https://doi.org/10.1016/j.patcog.2017.07.030
4. Daller, É., Bougleux, S., Gaüzère, B., Brun, L.: Approximate graph edit distance by several local searches in parallel. In: International Conference on Pattern Recognition Applications and Methods, pp. 149–158 (2018). https://doi.org/10.5220/0006599901490158
5. Ferrer, M.: Theory and algorithms on the median graph. Application to graph-based classification and clustering. Ph.D. thesis, Universitat Autònoma de Barcelona (2008). http://hdl.handle.net/10803/5788
6. Ferrer, M., Bardají, I., Valveny, E., Karatzas, D., Bunke, H.: Median graph computation by means of graph embedding into vector spaces. In: Fu, Y., Ma, Y. (eds.) Graph Embedding for Pattern Analysis, pp. 45–71. Springer, New York (2013). https://doi.org/10.1007/978-1-4614-4457-2_3
7. Ferrer, M., Karatzas, D., Valveny, E., Bardaji, I., Bunke, H.: A generic framework for median graph computation based on a recursive embedding approach. Comput. Vis. Image Underst. 115(7), 919–928 (2011). https://doi.org/10.1016/j.cviu.2010.12.010
8. Ferrer, M., Valveny, E., Serratosa, F., Riesen, K., Bunke, H.: Generalized median graph computation by means of graph embedding in vector spaces. Pattern Recogn. 43(4), 1642–1655 (2010). https://doi.org/10.1016/j.patcog.2009.10.013
9. Hlaoui, A., Wang, S.: Median graph computation for graph clustering. Soft Comput. 10(1), 47–53 (2006). https://doi.org/10.1007/s00500-005-0464-1
10. Jiang, X., Munger, A., Bunke, H.: On median graphs: properties, algorithms, and applications. IEEE Trans. Pattern Anal. Mach. Intell. 23(10), 1144–1151 (2001). https://doi.org/10.1109/34.954604
11. Moreno-García, C.F., Serratosa, F., Jiang, X.: Correspondence edit distance to obtain a set of weighted means of graph correspondences. Pattern Recogn. Lett. (2018). https://doi.org/10.1016/j.patrec.2018.08.027
12. Mukherjee, L., Singh, V., Peng, J., Xu, J., Zeitz, M.J., Berezney, R.: Generalized median graphs and applications. J. Comb. Optim. 17(1), 21–44 (2009). https://doi.org/10.1007/s10878-008-9184-7
13. Musmanno, L.M., Ribeiro, C.C.: Heuristics for the generalized median graph problem. Eur. J. Oper. Res. 254(2), 371–384 (2016). https://doi.org/10.1016/j.ejor.2016.03.048
14. Nienkötter, A., Jiang, X.: Improved prototype embedding based generalized median computation by means of refined reconstruction methods. In: Robles-Kelly, A., Loog, M., Biggio, B., Escolano, F., Wilson, R. (eds.) S+SSPR 2016. LNCS, vol. 10029, pp. 107–117. Springer, Cham (2016). https://doi.org/10.1007/978-3-319-49055-7_10

15. Rebagliati, N., Solé-Ribalta, A., Pelillo, M., Serratosa, F.: On the relation between the common labelling and the median graph. In: Gimel'farb, G., et al. (eds.) SSPR /SPR 2012. LNCS, vol. 7626, pp. 107–115. Springer, Heidelberg (2012). https:// doi.org/10.1007/978-3-642-34166-3_12
16. Riesen, K.: Structural Pattern Recognition with Graph Edit Distance. ACVPR. Springer, Cham (2015). https://doi.org/10.1007/978-3-319-27252-8

An Hypergraph Data Model for Expert Finding in Multimedia Social Networks

Vincenzo Moscato[1,2], Antonio Picariello[1,2], and Giancarlo Sperlí[1,2(⊠)]

[1] University of Naples "Federico", via Claudio 21, 80125 Naples, Italy
{vincenzo.moscato,antonio.picariello,giancarlo.sperli}@unina.it
[2] CINI - ITEM National Lab, via Cinzia, Complesso Universitario Montesantangelo, 80125 Naples, Italy

Abstract. Nowadays, the tremendous usage of multimedia data within *Online Social Networks* (OSNs) has led the born of a new generation of OSNs, called *Multimedia Social Networks* (MSNs). They represent particular social media networks – particularly interesting for *Social Network Analysis* (SNA) applications – that combine information on users, belonging to one or more social communities, together with all the multimedia contents that can be generated and used in the related environments. In this work, we present a novel *expert finding* technique exploiting a hypergraph-based data model for MSNs. In particular, some user *ranking* measures, obtained considering only particular useful hyper-paths, have been profitably used to evaluate the related expertness degree with respect to a given social topic. Several preliminary experiments on Last.fm show the effectiveness of the proposed approach, encouraging the future work in this direction.

Keywords: Multimedia Social Networks · Social Network Analysis · Expert finding · Hypegraphs

1 Introduction

In the last decades, the tremendous diffusion of *Online Social Networks* (OSNs) has introduced a new communication paradigm, where by now people around the world interact among them and spread/share any kind of information for different purposes within large user communities using Internet-enabled social networking applications (e.g., Facebook, Instagram, Last.fm, Twitter, Flickr, TripAdvisor, YELP, etc.).

Currently, there is not a generally one well-established definition for OSNs. Schneider et al. [6] define OSNs as users' communities composed by people that share common interest, activities, backgrounds, and/or friendships and can interact with others in numerous ways, directly or by means of posted information. Other definitions consider OSNs as a new particular type of virtual community [3] or advanced social networking software applications [5], heightening respectively social or technological innovatory aspects at the basis of their utilization and diffusion.

© Springer Nature Switzerland AG 2019
D. Conte et al. (Eds.): GbRPR 2019, LNCS 11510, pp. 110–120, 2019.
https://doi.org/10.1007/978-3-030-20081-7_11

In addition, the fast advance of the Information and Communication Technologies has enhanced OSN features, enabling users to share their lives, behaviors, works and interests by increasingly using multimedia items (e.g., text, audio, video, image) and to interact with such objects in order to provide feedbacks, comments, opinions or feelings with respect to the posted data.

This phenomenon has led to the born of a new type of OSN, called *Multimedia Social Network* (MSN) [1,2]. It can be seen as a particular social media network that combines information on users, belonging to one or more social communities, together with all the multimedia contents (such as music, images, posts, etc.) that can be generated and used within the related environments.

A lot of classical *Social Network Analysis* (SNA) applications, including influence analysis, social recommendation, viral marketing, event recognition, expert finding, community detection, user profiling, social data privacy and so on, can effectively take advantage of MSNs' characteristics as the different kinds of relationships (i.e., among multimedia contents, among users and multimedia content, in addition to those among users themselves) that are typical of such networks or, furthermore, multimedia features of shared objects [1,2].

In this work, we present a novel *expert finding* technique within social networks exploiting a hypergraph-based data model for MSNs. In particular, the hypergraph model permits to easily represent all the described MSNs' relationships and to enable several kinds of analytics, such as finding experts, by means of the introduction of some *ranking* criteria. User ranking measures, obtained considering only particular useful hyperpaths, can be profitably adopted to evaluate the expertness degree with respect to a given social topic. We also propose a strategy for hypergraph building from data coming from different OSNs (e.g. Facebook, Twitter, Flickr, Instagram, Last.fm, Youtube).

The paper is organized as the following. Section 2 discusses the related work concerning expert finding within social networks. Section 3 describes the adopted MSN data model together with the related ranking/centrality measures and the hypergraph building process. Section 4 outlines the expert finding system architecture with some implementation details. Finally, Sect. 5 reports some experiments using a dataset from Last.fm and provides conclusions and future work.

2   Related Work

The large amount of heterogeneous data shared on Multimedia Social Networks requires novel techniques and methodologies to support users in addressing their needs.

Concerning social networks having as main purpose to share comment and opinions on specific topics/items, users' choice for a given object is made difficult by the high number and heterogeneity of reviews. Thus, identifying users having knowledgeable in specified areas, also called *experts*, is a critical challenge for a lot of applications.

The existing *expert finding* approaches can be classified into two main groups: *authority-oriented* and *topic-oriented*.

The approaches of first family are mainly focused on link analysis on whom users are ranked. In [4] the authors propose a technique based on *HyperLink Induced Topic Search* (HITS) to classify users in spammers and experts within Twitter on the basis of some features of tweets. Zhu et al. [13] introduce an expert finding framework based on link analysis approach for ranking user authority by considering information belonging to relevant knowledge categories. Another methodology based on chains of social referrals and profile matching on local information in OSNs has been described by [9]. Furthermore, an interesting approach has been detailed by Zhao et al. [10] using a ranking learning metric that combines users relative quality rank w.r.t. given questions and their social relations. Finally, Zhao et al. [11] propose a graph-regularized matrix completion algorithm for inferring user ranking employing also information from social networks to deal with missing values problem.

In turn, the second family approaches are mainly focused on latent topic modelling techniques. A topic-sensitive probabilistic model has been described in [12] combining link and user analysis for identifying experts in Question Answer Communities. A system for expert finding, namely *BMExpert*, has been the proposed by Wang et al. [7] combining relevance of documents with respect to the query topic, importance of documents, and associations between documents and experts. Finally, Wei et al. [8] introduce a semi-supervised graph-based ranking approach that combines three different types of relations (between users, lists and users and list) for computing global authority of users.

The proposed approach combines some characteristics of topic and link analysis based techniques, leveraging also multimedia objects' features for ranking user authority about a given topic. In particular, the proposed approach is based on a data model relying on hypergraph data structure to represent the well-known heterogeneous relationships of MSN and exploits some ranking criteria obtained considering only useful MSN hyperpaths.

3 Modeling Multimedia Social Networks

3.1 Basic Definitions

In our vision, a MSN is basically composed by three different entities:

– **Users** - the set of persons and organizations constituting the particular social community. Several information concerning their profile, interests, preferences, etc. can be opportunely considered in our model.
– **Objects** - the user-generated multimedia content that is of interest within a given social community. Object can be obviously described using *metadata* (eventually correlated to some annotation schemata) and low-level *features*.
– **Topics** - the most significant terms or named entities of one or more domains, exploited by users to describe multimedia items and eventually derived from the analysis of textual information, mainly tags but also keywords, labels, comments, reviews etc.

Several types of relationships can be established among the described entities: for example, a user can annotate an image with a particular tag, two friends can comment the same post, a user can listen a song, a user can share some videos within a group and so on. Due to the variety and intrinsic complexity of these relationships, we decided to leverage the *hypergraph* formalism to model an MSN. In the following, we provide all the basic definitions necessary to characterize our model.

Definition 31 (MSN). *An* Multimedia Social Network *MSN is a triple* $(V; E = \{e_i : i \in I\}; \omega)$, *V being a finite set of* vertices, *E a set of* hyperedges *with a finite set of indexes I and* $\omega : E \to [0, 1]$ *a weight function. The set of vertices is defined as* $V = U \cup O \cup T$, *U being the set of MSN users, O the set of multimedia objects and* T *the set of topics. Each hyperedge* $e_i \in e$ *is in turn defined by a ordered pair* $e_i = (e_i^+ = (V_{e_i}^+, i); e_i^- = (i, V_{e_i}^-))$. *The element* e_i^+ *is called the* tail *of the hyperarc* e_i *whereas* e_i^- *is its* head, $V_{e_i}^+ \subseteq V$ *being the set of vertices of* e_i^+, $V_{e_i}^- \subseteq V$ *the set of vertices of* e_i^- *and* $V_{e_i} = V_{e_i}^+ \cup V_{e_i}^-$ *the subset of vertices constituting the whole hyperedge.*

An hypergraph related to a MSN can be also denoted by an *incidence matrix* H with entries as:

$$h(v, e_i) = \begin{cases} 1, & \text{if } v \in V_{e_i} \\ 0, & \text{otherwise} \end{cases} \tag{1}$$

Actually, vertices and hyperedges are *abstract data types* and we use the "dot notation" to identify the attributes of a given vertex or hyperedge: as an example, e_i.id, e_i.name, e_i.time, e_i.source and e_i.type represent the id, name, timestamp, source (social network) and type of the hyperedge e_i, respectively. In turn, the weight function can be used to define the "confidence" of a given relationship.

Definition 32 (Social Path). *A social path between vertices* v_{s_1} *and* v_{s_k} *of an MSN is a sequence of distinct vertices and hyperedges* $v_{s_1}, e_{s_1}, v_{s_2}, ..., e_{s_{k-1}}, v_{s_k}$ *such that* $\{v_{s_i}, v_{s_{i+1}}\} \subseteq V_{e_{s_i}}$ *for* $1 \leq i \leq k - 1$. *The length of the hyperpath is* $\alpha \cdot \sum_{i=1}^{k-1} \frac{1}{\omega(e_{s_i})}$, α *being a normalizing factor. We say that a social path contains a vertex* v_h *if* $\exists e_{s_i} : v_h \in e_{s_i}$.

Social paths between two nodes can "directly" or "indirectly" connect two users because they are "friends" or as they commented the same video.

Definition 33 (Distances). *We define* minimum distance $(d_{min}(v_k, v_j))$, *maximum distance* $(d_{max}(v_k, v_j))$ *and average distance* $(d_{avg}(v_k, v_j))$ *between two vertices of an MSN the length of the shortest hyperpath, the length of the longest hyperpath and the average length of the hyperpaths between* v_k *and* v_j, *respectively. In a similar manner, we define the* minimum distance $(d_{min}(v_k, v_j | v_z))$, *maximum distance* $(d_{max}(v_k, v_j | v_z))$ *and average distance* $(d_{avg}(v_k, v_j | v_z))$ *between two vertices* v_k *and* v_j, *for which there exists a hyperpath containing* v_z.

Definition 34 (λ-Nearest Neighbors Set). *Given a vertex $v_k \in V$ of an MSN, we define the λ-Nearest Neighbors Set of v_k the subset of vertices NN_k^λ such that $\forall v_j \in NN_k^\lambda$ we have $d_{min}(v_k, v_j) \leq \lambda$ with $v_j \in U$. Considering only the constrained hyperpaths containing a vertex v_z, we denote with NN_{iz}^λ the set of nearest neighbors of v_k such that $\forall v_j \in NN_{iz}^\lambda$ we have $\bar{d}_{min}(v_k, v_j | v_z) \leq \lambda$.*

If we consider as neighbors only vertices belonging to user type, the NN^λ set is called λ-*Nearest Users Set* and denoted as NNU^λ, similarly in case of objects we define the λ-*Nearest Objects Set* as NNO^λ.

3.2 Relationships

We have identified three different categories:

- **User to User** relationships, describing user actions towards other users;
- **Similarity** relationships, describing a relatedness between two objects, users or topics;
- **User to Object** relationships, describing user actions on objects, eventually involving some topics or other users.

In the following, we provide the definition for each class of relationship.

Definition 35 (User to User relationship). *Let $\widehat{U} \subseteq U$ a subset of users in a MSN, we define* user to user relationship *each hyperedge e_i with the following properties:*

1. *$V_{e_i}^+ = u_k$ such that $u_k \in \widehat{U}$,*
2. *$V_{e_i}^- \subseteq \widehat{U} - u_k$.*

Examples of "user to user" relationships are properly represented by *friendship*, *following* or *membership* of some online social networks.

Definition 36 (Similarity relationship). *Let $v_k, v_j \in V$ ($k \neq j$) two vertices of the same type of a MSN, we define* similarity relationship *each hyperedge e_i with $V_{e_i}^+ = v_k$ and $V_{e_i}^- = v_j$. The weight function for this relationship returns similarity value between the two vertices.*

The similarity relationships are defined on the top of a *similarity function* $f_{sim} : V \times V \rightarrow R$. It is possible to compute a similarity value: (i) between two users by considering different types of features (interests, profile information, preferences, etc.); (ii) between two objects using the well-known (high and low level) features and metrics proposed in the literature; (iii) between two annotation assets exploiting the related topics and the well-known metrics on vocabularies or ontologies.

In our model, a similarity hyperedge is effectively generated if $\omega(e_i) \geq \theta$, θ being a given threshold.

Definition 37 (User to Object relationship). *Let $\widehat{U} \subseteq U$ a set of users in a MSN and $\widehat{O} \subseteq O$ a set of objects, we define* user to multimedia relationship *each hyperedge e_i with the following properties:*

1. $V_{e_i}^+ = u_k$ *such that* $u_k \in \widehat{U}$,
2. $V_{e_i}^- \subseteq \widehat{O} \cup T$.

Examples of "user to object" relationships are represented, as an example, by *publishing, reaction, annotation, review, comment* (in the last three cases the set $V_{e_i}^-$ also contains one or more topics) or *user tagging* (involving also one ore more users) activities.

3.3   Hypergraph Building

The proposed hypergraph building process consists of three different stages: (i) hypergraph structure construction; (ii) topic distribution; (iii) similarity learning. First, extracted data related to relationships between objects, users and objects and users themselves are initially used to construct the hypergraph structure in terms of nodes and hyperedges.

For the user to object relationships, textual annotations are then analyzed by the LDA approach to learn the most important topics and to infer relations between topics and textual annotations. From the other hand, similarity values between users, objects and topics are eventually determined using proper strategies.

3.4   Centrality Measures for Expert Finding

One of the fundamentals in Social Network Analysis is to compute the *centrality* measures for ranking user nodes of a social graph. As well known, the centrality represents the "importance" of a given user within the related community and can be easily exploited for several applications.

Here, we propose to use centrality as the main measure for detecting experts in a certain domain, analyzing Multimedia OSN.

In the literature, there exists a lot of measures to determine the centrality of a node in a social graph. In this work, we define a new centrality measure based on the concept of "neighborhood" among users through λ-Nearest Neighbors Set in MSNs. In particular, the novelty of this centrality measure concerns the use of multimedia object and annotation asset for defining new paths to connect two users

Definition 38 (Neighborhood Centrality). *Let $v_k \in V$ a vertex of a MSN and λ a given threshold, we define the* neighborhood centrality *of v_k as:*

$$nc(v_k) = \frac{|NN_{v_k}^\lambda \cap V|}{|V| - 1} \tag{2}$$

NN_k^λ *being the λ-Nearest Users Set of v_k.*

The neighborhood centrality of a given node can be measured by the number of nodes that are "reachable" within a certain number of steps using social paths. The introduced measure can be computed locally with respect to a community of users ($\widehat{U} \subseteq U \subseteq V$) and considering only vertices of user type for the end-to-end nodes of hyperpaths. In this manner, centrality is refereed to user importance within the related community. We define *user centrality* such kind of measure.

In addition, in order to give more importance to user-to-content relationships during the computation of distances for the user neighborhood centrality, we can apply a *penalty* if the considered hyperpaths contain some users; in this way, all the distances can be computed as $\tilde{d}(v_k, v_j) = d(v_k, v_j) + \beta \cdot N$, N being the number of user vertices in the hyperpath between v_i and v_j and β a scaling factor. This strategy has been chosen because an expert, in our opinion, is defined according to its behavior on MSN describing by published multimedia object and annotation asset.

Eventually, we can obtained a *topic-sensitive user neighborhood centrality* considering in the distances' computation only hyperpaths that contain a given topic node:

$$nc(u_k|\widehat{U}, t_z) = \frac{\left| NNU_{u_k t_z}^{\lambda} \cap \widehat{U} \right|}{\left| \widehat{U} \right| - 1} \tag{3}$$

$\widehat{U}$ being a user community, u_k a single user and t_z a given topic.

4   System Architecture

Figure 1 shows an overview of the proposed system architecture. More in detail, it can be divided into the following three main layers: *Data Ingestion, Knowledge Management* and *Social Network Analysis* that provides expert finding tools.

In the first layer data coming from heterogeneous OSN (such as Facebook, Twitter, LastFM etc.) are crawled by using their owner API and stored into a No-SQL columnar database Cassandra[1] for properly storing large amount of data.

The *Knowledge management* layer has the aim to extract information from the No-SQL database for building the MSN data model (Hypergraph Building Module) and storing it into HypergraphDB[2], a No-SQL database based on hypergraph data structure.

Eventually, the *Social Network Analysis* layer is composed by the *Expert Finding* module relying on HyperX[3], a framework built upon Apache Spark[4] for processing hypergraph, to rank users using centrality with respect to a given topic and *Visualization* module, based on Jung API[5], to represent and provide insights about the analyzed network.

---

[1] http://cassandra.apache.org/.
[2] http://www.hypergraphdb.org/.
[3] https://github.com/jinhuang/hyperx.
[4] https://spark.apache.org/.
[5] http://jung.sourceforge.net/.

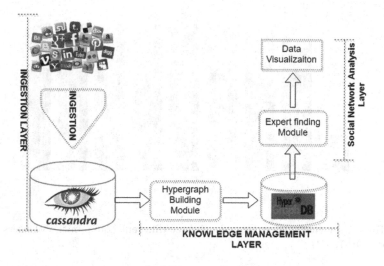

Fig. 1. Architectural overview.

5   Results and Discussion

In this section we describe the preliminary experiments for evaluating the effectiveness of the proposed approach. The evaluation has been made on a music collection[6] containing a set of data extracted from the Last.fm multimedia social network in the first half of 2009. Table 1 shows in detail the MSN characterization.

Table 1. MSN characterization

Element	Number
Crawled user	99,405
Annotations	10,936,545
Items	1,393,559
Tags	281,818
Groups	66,429

The set of nodes is composed by users, topics unveiled by the analysis of tags in the annotation and multimedia items; hyperarc set is in turn composed by the following relationships: friendship, membership, annotation and user and multimedia similarity. In particular, the similarity between two users has been

---

[6] http://carl.cs.indiana.edu/data/last.fm/.

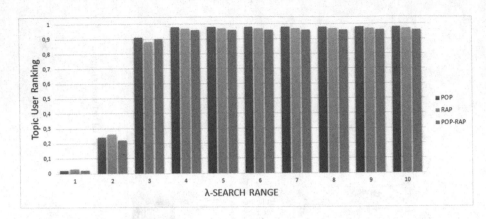

Fig. 2. Topic User Ranking computed on three different communities of users (POP, RAP and POP-RAP).

Table 2. Ranking comparison (PageRank (PR), K-Step Markov (KS), MSN Topic User Ranking (MSNTUR), Human Ranking (HR)).

	τ	ρ
MSNTUR - PR	0,48	0,58
MSNTUR - KS	0,65	0,78
MSNTUR - HR	0,80	0,91
PR - HR	0,70	0,74
KS - HR	0,67	0,82

computed according to neighborhood values provided by the Last.fm API whilst Spotalike[7] facilities (based on the low-level features) together with Last.fm similarity score is used for evaluating similarity between two songs.

Figure 2 shows the average values of users ranking for each community varying λ. As easy to know, the ranking value for each user assumes the same value when λ's value grows up.

Thus, we compare the proposed ranking method based on the Neighborhood Centrality, choosing $\lambda = 2$, some well-known approaches and a human-generated ranking (representing the unique gold standard[8]) of users within Pop, Rap and Pop-Rap communities. Table 2 shows the obtained results in terms of *Kendall' Tau* (τ) and *Spearman's Rank Correlation* (ρ) coefficients.

Finally, Fig. 3 shows an example of the visualization module.

---

[7] http://www.spotalike.com/.

[8] We ask a group of our students to rank the users expertness w.r.t. the different communities considering number and relevance and of the related comments.

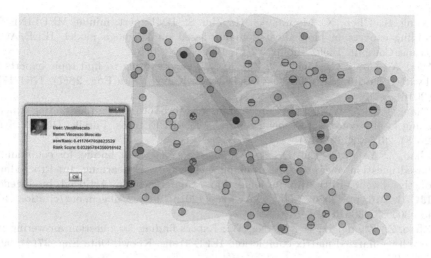

Fig. 3. Example of the visualization module.

6   Conclusion

In this paper we propose an expert finding technique based on hypergraph data model in which multimedia objects and annotation assets plays a key role. The obtained results shows the goodness of the approach in detection expert finding w.r.t. human ground truth and encourage the future work in this direction.

References

1. Amato, F., Moscato, V., Picariello, A., Sperlí, G.: Multimedia social network modeling: a proposal. In: 2016 IEEE Tenth International Conference on Semantic Computing (ICSC), pp. 448–453. IEEE (2016)
2. Amato, F., Moscato, V., Picariello, A., Sperlí, G.: Diffusion algorithms in multimedia social networks: a preliminary model. In: Proceedings of the 2017 IEEE/ACM International Conference on Advances in Social Networks Analysis and Mining 2017, pp. 844–851. ACM (2017)
3. Dwyer, C., Hiltz, S., Passerini, K.: Trust and privacy concern within social networking sites: a comparison of Facebook and MySpace. In: AMCIS 2007 Proceedings, p. 339 (2007)
4. Khan, M.U.S., Ali, M., Abbas, A., Khan, S.U., Zomaya, A.Y.: Segregating spammers and unsolicited bloggers from genuine experts on Twitter. IEEE Trans. Dependable Secure Comput. **15**(4), 551–560 (2018)
5. Richter, D., Riemer, K., vom Brocke, J.: Internet social networking. Wirtschaftsinformatik **53**(2), 89–103 (2011)
6. Schneider, F., Feldmann, A., Krishnamurthy, B., Willinger, W.: Understanding online social network usage from a network perspective. In: Proceedings of the 9th ACM SIGCOMM Conference on Internet Measurement Conference, pp. 35–48. ACM (2009)

7. Wang, B., Chen, X., Mamitsuka, H., Zhu, S.: BMExpert: mining MEDLINE for finding experts in biomedical domains based on language model. IEEE/ACM Trans. Comput. Biolo. Bioinf. **12**(6), 1286–1294 (2015)
8. Wei, W., Cong, G., Miao, C., Zhu, F., Li, G.: Learning to find topic experts in Twitter via different relations. IEEE Trans. Knowl. Data Eng. **28**(7), 1764–1778 (2016)
9. Zhang, L., Li, X.Y., Lei, J., Sun, J., Liu, Y.: Mechanism design for finding experts using locally constructed social referral web. IEEE Trans. Parallel Distrib. Syst. **26**(8), 2316–2326 (2015)
10. Zhao, Z., Yang, Q., Cai, D., He, X., Zhuang, Y.: Expert finding for community-based question answering via ranking metric network learning. In: Proceedings of the Twenty-Fifth International Joint Conference on Artificial Intelligence, IJCAI 2016, pp. 3000–3006. AAAI Press (2016). http://dl.acm.org/citation.cfm?id=3060832.3061041
11. Zhao, Z., Zhang, L., He, X., Ng, W.: Expert finding for question answering via graph regularized matrix completion. IEEE Trans. Knowl. Data Eng. **27**(4), 993–1004 (2015)
12. Zhou, G., Zhao, J., He, T., Wu, W.: An empirical study of topic-sensitive probabilistic model for expert finding in question answer communities. Knowl.-Based Syst. **66**, 136–145 (2014)
13. Zhu, H., Chen, E., Xiong, H., Cao, H., Tian, J.: Ranking user authority with relevant knowledge categories for expert finding. World Wide Web **17**(5), 1081–1107 (2014)

On-Line Learning the Edit Costs Based on an Embedded Model

Elena Rica$^{(\boxtimes)}$, Susana Álvarez, and Francesc Serratosa

Universitat Rovira i Virgili, Tarragona, Spain
{mariaelena.rica,susana.alvarez,francesc.serratosa}@urv.cat

Abstract. This paper presents an on-line learning method to automatically deduce the insertion, deletion and substitution costs of the graph edit distance, which is inspired in a previously published off-line learning method. The original method is based on embedding the ground-truth node-to-node mappings into a Euclidean space and learning the edit costs through the hyper-plane in this new space that splits the nodes into the mapped ones and the non-mapped ones. The new method has the advantage of learning the edit costs and computing the graph edit distance can be done simultaneously. Experimental validation shows that the matching accuracy is competitive with the off-line method but without the need of the whole learning set.

Keywords: On-line learning · Graph edit distance ·
Learning edit costs · Graph-matching algorithm

1 Introduction

Attributed relational graphs are commonly used as abstract representations for common structures such as documents, images or chemical compounds, among others [12]. Nodes of graphs represent local parts of the object and edges represent the relations between these local parts. Error-tolerant graph matching algorithms [3,19] are applied to deduce the distance between prototypes represented by attributed graphs. Error-tolerant graph matching algorithms are based on finding a mapping between nodes so that both graphs look similar when their nodes are mapped according to this node-to-node mapping. One of the most used frameworks to define the error-tolerant graph matching is through the graph edit distance [6,16,18]. The main idea is to define the difference between graphs as the amount of distortion required to transform one graph into another through substituting, deleting or inserting nodes and edges. To do so, some penalty costs are defined for these edit operations.

In this paper, we present an on-line method to learn the graph edit distance costs. The aim is to recompute these values automatically when new data is available. Thus, in a recognition process, the graph edit distance can be computed having these costs as input parameters, which have been found trough our optimisation process. In this way, the recognition process can be carried out at the same time than the learning process. The method is inspired by an off-line learning algorithm recently published [1]. Nevertheless, our algorithm is not a simple iterating process on the off-line algorithm, but some processes have been incorporated to keep the algorithm learning with the minimum data.

Note that, in some object retrieval applications, in which elements are represented by graphs, the aim is to deduce which are the most similar graphs, without the graphs being previously classified. In these cases, it is crucial the learning method to learn the edit costs such that the best node-to-node mapping between pairs of graphs is computed instead of maximising the classification ratio. This is the reason why the whole process we present is dependent on a ground-truth node-to-node matching. Recently, a graph database generator that returns pairs of graphs with their ground-truth correspondence has been presented [15].

To our knowledge, any on-line method has been published yet to learn the graph edit distance parameters, although there are several published off-line methods that learn them [1,2,4,5,7,10,11,17]. An important feature of these off-line methods is the type of costs the learning algorithm obtains: a self-organising map [10], a probability density function [11] or linear functions [1,2,4,5,7,17].

The main characteristic of the off-line methods is they learn with the whole data at once, however, the on-line methods receive subsets of data and make successive learning processes with them. We present an on-line method that learns the insertion and deletions costs (similarly to [1,4]) and the weights on the substitutions costs (similarly to [1,2,5,7]).

The outline of the paper is as follows. In the next section, we define attributed graphs and graph edit distance. In Sect. 3, we present our learning strategy. In Sect. 4, we show the experimental validation and finally, in Sect. 5, we conclude the article.

2   Graph Edit Distance

The graph edit distance (GED) between two attributed graphs is defined as the transformation from one graph into another, through the edit operations, that obtains the minimum cost. These edit operations are: substitution, deletion and insertion of nodes and also edges. Every edit operation has a cost depending on the attributes on the involved nodes or edges. This graph transformation can be defined through a node-to-node mapping f between nodes of both graphs.

Having a pair of graphs, G and G', a correspondence f between these graphs is a bijective function that assigns one node of G to only one node of G'. We suppose both graphs have the same number of nodes since they have been expanded with new nodes that have a concrete attribute. We call these new nodes as $Null$. Note that the mapping between edges is imposed by the mapping of the nodes whose edges are connected.

We define G_i as the ith node in G, G'_a as the ath node in G', $G_{i,j}$ as the edge between the ith node and the jth node in G, $G'_{a,b}$ as the edge between the ath node and the bth node in G'. Nodes and edges have N and M attributes, which are real numbers, respectively. Moreover, γ_i^t is the tth attribute of node G_i and

$\beta_{i,j}^t$ is the tth attribute of edge $G_{i,j}$. We also define the mapping $f(i) = a$ from G_i to G'_a, we say that it represents a node substitution if both nodes are not $Null$. Contrarily, if node G'_a is a $Null$ and G_i is not, we say that it represents a deletion. Finally, if node G_i is a $Null$ and G'_a is not, we say that it represents an insertion. Similarly happens with the edges. The case that both nodes or both edges are null is not considered since it is defined as the cost is always zero.

We define the GED as follows:

$$GED(G, G') = \min_{f:G\to G'} \left\{ \sum_{\forall G_i} C^n(i, f(i)) + \sum_{\forall G_{i,j}} C^e(i, j, f(i), f(j)) \right\} \quad (1)$$

Where, functions $C^n(i, f(i))$ and $C^e(i, j, f(i), f(j))$ represent the cost of mapping a pair of nodes (G_i and $G'_{f(i)}$) and a pair of edges ($G_{i,j}$ and $G'_{f(i),f(j)}$), respectively, and they are defined through the cost functions in Eqs. 2 and 3.

$$C^n(i, a) = \begin{cases} C_S^n(i, a) & \text{if } G_i \neq Null \wedge G'_a \neq Null \\ C_D^n(i) & \text{if } G_i \neq Null \wedge G'_a = Null \\ C_I^n(a) & \text{if } G_i = Null \wedge G'_a \neq Null \end{cases} \quad (2)$$

$$C^e(i, j, a, b) = \begin{cases} C_S^e(i, j, a, b) & \text{if } G_{i,j} \neq Null \wedge G'_{a,b} \neq Null \\ C_D^e(i, j) & \text{if } G_{i,j} \neq Null \wedge G'_{a,b} = Null \\ C_I^e(a, b) & \text{if } G_{i,j} = Null \wedge G'_{a,b} \neq Null \end{cases} \quad (3)$$

Where $C_S^n(i, a)$ is the cost of substituting node G_i by node G'_a, $C_D^n(i)$ is the cost of deleting node G_i and $C_I^n(a)$ is the cost of inserting node G'_a. Similarly, $C_S^e(i, j, a, b)$ is the cost of substituting edge $G_{i,j}$ by edge $G'_{a,b}$, $C_D^e(i, j)$ is the cost of deleting edge $G_{i,j}$ and $C_I^e(a, b)$ is the cost of inserting edge $G'_{a,b}$.

In this paper, we impose the restrictions $C_I^n(a) = C_D^n(i) = K^n$ and $C_I^e(i, j) = C_D^e(i, j) = K^e$, where K^n and K^e are real numbers. Moreover, $C_S^n(i, a) = \sum_{t=1}^{N} w_t^n \left\| \gamma_i^t - \gamma_a'^t \right\|$ and $C_S^e(i, j, a, b) = \sum_{t=1}^{M} w_t^e \left\| \beta_{i,j}^t - \beta_{a,b}'^t \right\|$, where $w^n = (w_1^n, ..., w_N^n)$ is the vector of nodes attributes' weights and $w^e = (w_1^e, ..., w_M^e)$ is the vector of edges attributes' weights. Furthermore,

$$1 = \sum_{t=1}^{N} w_t^n \qquad 1 = \sum_{t=1}^{M} w_t^e \quad (4)$$

3 Learning the Graph Edit Costs

In this section, we first plainly summarise the off-line method presented in [1] and then we move on to explain our on-line proposal. It is crucial to explain the off-line method since our method is inspired in it.

3.1   Off-Line Learning the Graph Edit Costs

The basic scheme of the off-line method is summarised in Fig. 1. The system receives a set of triplets composed of two graphs and a ground-truth correspondence between them, $\{(G, G', f)_1, (G, G', f)_2, ...\}$, and outputs the substitution weights on nodes and edges and also the deletion and insertion costs on nodes and edges. Figure 1 only shows one triplet composed of two graphs that have five and four nodes, respectively. The ground-truth correspondence is represented through the dashed arrows. Four nodes are substituted and one node is deleted.

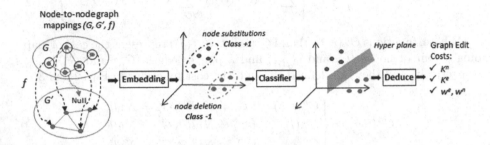

Fig. 1. Basic scheme of the off-line learning method.

The algorithm is composed of three main steps:

In the first step (Embedding), the ground-truth node-to-node mappings are embedded into a Euclidean space S, being $S = (S_1^n, ..., S_N^n, S_1^e, ..., S_M^e, S_{K^e})$ of dimension $N + M + 1$. Each node substitution is transformed into a point in this space and it is assigned to the "+1" class. Moreover, each node deletion is transformed into $\tilde{N}$ points, which are assigned to the "−1" class. $\tilde{N}$ is the number of substituted nodes in the ground-truth correspondence. The ground-truth correspondence in Fig. 1 makes the embedding step to generate four points that represent the four node substitution operations (one point per substitution) and four points that represent the only one node deletion (the number of points that generate each deletion is the number of substituted nodes).

In the second step (Classifier), a linear hyper-plane is computed that has to be the best linear border between both classes. Authors in [1] describe that any known linear classifier that return the hyper-plane can be used. Equation 5 defines this border, as described in [1]. Note the constants in this hyper-plane are the substitution weights $w_2^n, ..., w_N^n$ and $w_2^e, ..., w_M^e$ and also the insertion and deletion costs on nodes and edges K^n and K^e, respectively. Finally, note that w_1^n and w_1^e do not appear in Eq. 5.

$$S_1^n + w_2^n \cdot S_2^n + ... + w_N^n \cdot S_N^n + S_1^e + w_2^e \cdot S_2^e + ... + w_M^e \cdot S_M^e + \\ K^e \cdot S_{K^e} + K^n = 0 \tag{5}$$

For explanatory reasons, Fig. 2 shows the specific case of $N = M = 1$, where S is a 3D dimensional space. In this example, graphs have three and two nodes (not shown in the figure). The ground-truth correspondence imposes two nodes to be substituted (they generate two points) and one node to be deleted (that also generates two points).

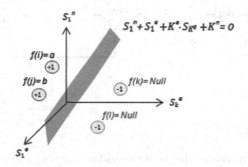

Fig. 2. Hyper-plane obtained in the learning process when M = N = 1.

Finally, in the last step (Deduce), weights $w_2^n, ..., w_N^n$ and $w_2^e, ..., w_M^e$ and also constants K^n, and K^e are extracted from the hyper-plane constants. Moreover, w_1^n and w_1^e are obtained through Eq. 4.

3.2   On-Line Learning the Graph Edit Costs

While the off-line method embeds all the set of triplets at once, the on-line method receives a triplet (G, G', f) at a time and computes K^n, K^e, w^n and w^e each time it is executed. To do so, it is needed the algorithm to input not only (G, G', f) but also two sets of embedded points D_{-1} and D_1 that have been computed in the previous execution (they are empty sets in the first iteration of the algorithm). The algorithm is composed of four main steps (Fig. 3):

Embedding (G, G', f) into the Euclidean space S is the first step of the algorithm (Line 1) and it is done in the same way that the off-line method [1] does. More precisely, it generates two sets, *new* D_1 and *new* D_{-1}, composed of points in the space S that represent the node substitutions and node deletions in (G, G', f), respectively.

Feeding (Lines 2 and 3) is a simple process in which the new sets *new* D_1 and *new* D_{-1} and the previous ones D_1 and D_{-1} (which are input parameters of the algorithm) are put together.

Then, Data reduction (Lines 4 to 14) updates sets D_1 and D_{-1} with the aim of reducing the amount of points but holding two main properties of these sets. The first one is keeping the general distance between points as well as their positions. This means that we want to have less points but maintain the same information of the sets as much as possible. The second property is keeping the same relation of the number of points of both sets. This is because, all the classifiers are biased by the order of the sets. In this case, we want to keep the number of points proportion to be as much reliable as possible to the input data. The input parameter k^{means} is the maximum number of points that will have

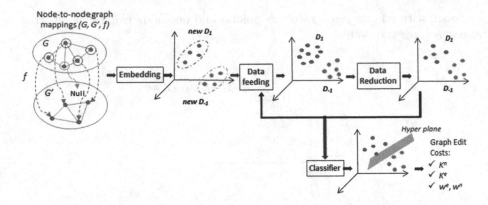

Fig. 3. Basic scheme of the on-line learning method.

the sets when each iterations finishes and the algorithm returns the graph edit distance parameters. From Line 4 to Line 11, the algorithm decides the number of elements that the updated sets D_1 and D_{-1} will have. Finally, in Lines 12 and 13, the reduction is done in each set. Note that we have selected the K-means clustering method [8] to perform this reduction although other reduction algorithms could be explored. Note the generated sets D_1 and D_{-1} are returned by the algorithm and feed the next iteration.

Finally, in the Classifier step (Lines 15 and 16), $w_2^n, ..., w_N^n, w_2^e, ..., w_M^e, K^n$ and K^e are extracted from the hyper-plane constants that a linear classifier returns as the border of sets D_1 and D_{-1}. Moreover, w_1^n and w_1^e are obtained through Eq. 4.

On-line Algorithm

Input$(K^{means}, D_1, D_{-1}, (G, G', f)) \rightarrow$ **Output**$(K^n, K^e, w^n, w^e, D_1, D_{-1})$

1. $(new\ D_1, new\ D_{-1}) = \text{Embedding}(G, G', f)$
2. $D_1 = D_1 \cup new\ D_1$
3. $D_{-1} = D_{-1} \cup new\ D_{-1}$
4.   **If** $|D_1| > K^{means} \vee |D_{-1}| > K^{means}$
5.   **If** $|D_1| \geq |D_{-1}|$
6. $K_1^{means} = K^{means}$
7. $K_{-1}^{means} = K^{means} * (|D_{-1}| / |D_1|)$
8.   **Else**
9. $K_1^{means} = K^{means} * (|D_1| / |D_{-1}|)$
10. $K_{-1}^{means} = K^{means}$
11.   **End if**
12. $D_1 = k\text{-}means(D_1, K_1^{means})$
13. $D_{-1} = k\text{-}means(D_{-1}, K_{-1}^{means})$
14.   **End if**
15. $[K^n, K^e, w_2^n, ..., w_N^n, w_2^e, ..., w_M^e] = \text{Classifier}\ (D_1, D_{-1})$
16. $w_1^n = 1 - \sum_{t=2}^{N} w_t^n \quad w_1^e = 1 - \sum_{t=2}^{M} w_t^e$
End Algorithm

4   Experimental Validation

The method we present has been tested using the Tarragona-Graph repository detailed in [9]. This repository has the main characteristic that each register is composed of a pair of graphs, a ground-truth correspondence between them (mapping between their nodes), and their class. It contains three graph databases: Letter_Low, Letter_Med and Letter_High that represent artificially distorted letters of the Latin alphabet with an increasing level of distortion. In each data base we have used a set of $37,500$ pairs of graphs for learning and a different set of $37,500$ pairs of graphs for testing. Every set generated $150,000$ points in the embedding space. We used the matching algorithm [13,14].

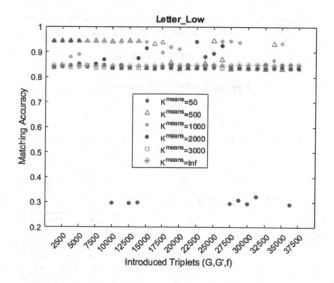

Fig. 4. Letter_Low accuracy.

The average matching accuracy obtained in the three data sets is shown in Figs. 4, 5 and 6, given a number of introduced triplets (G, G', f) taken from the test set and also different values of K^{means}. The value $K^{means} = Inf$ represents no reduction of the data, which is the same than applying the off-line algorithm [1], given the specific number of introduced triplets.

In Letter_Low (Fig. 4), accuracies generated by different values of K^{means} are stable and almost similar except for the ones generated by $K^{means} = 50$. Nevertheless, in Letter_Med (Fig. 5) and Letter_High (Fig. 6), the stability is achieved at $K^{means} = 2,000$ and $K^{means} = 3,000$, respectively. The off-line method ($K^{means} = Inf$) becomes to be the most stable. Note that, higher is the number of points K^{means} allowed in sets D_1 and D_{-1}, slower is the algorithm (see Table 1). Thus, we wish to keep this value as lower as possible. Nevertheless, we observe that this is a parameter that depends on the level of noise of the

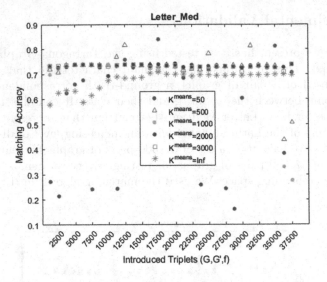

Fig. 5. Letter_Med accuracy.

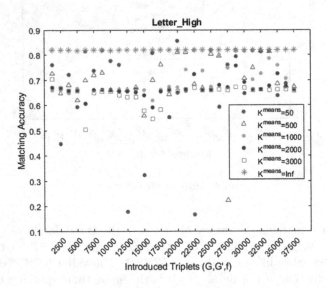

Fig. 6. Letter_High accuracy.

databases. Comparing our method to the off-line one [1], we realise that we achieve competitive accuracies, although having a huge data reduction. Note that in these databases, the off-line method generates 150,000 points but the on-line algorithm only needs 3,000 points, in the worst case (Letter_High).

Table 1. Run time in seconds given several K^{means}

K^{means}	50	500	1000	2000	3000	Off-line
Letter_Low	1.3	6.5	11.7	29.8	52.3	51.7
Letter_Med	1.3	5.5	11.1	25.6	68.4	52.1
Letter_High	1.2	6.6	12.0	25.7	67.1	55.4

5   Conclusions

We have presented an on-line method to learn the edit costs based on embedding the ground-truth correspondence into a Euclidean space. This space, which was previously defined in an off-line method, has the particularity that the border between node substitutions and deletions is set as a hyper-plane defined by the edit cost parameters. The learning method is limited to the applications that substitution costs are represented as a normalised euclidean distance and insertion and deletion costs are constants. Note that the weights and costs deduced through our algorithm do not guarantee to be the optimal ones in an optimal graph-matching algorithm. Each time our method is executed, the weights and edit costs are returned and also some embedded points. These points and the new node-to-node mappings are the input of the next algorithm iteration. From a practical point of view, our method has three main advantages. First, the learned weights and costs can be used each time the algorithm is computed. Second, only parameter K^{means} has to be tuned. And third, the graph edit distance does not need to be computed in the learning process, as it is needed in other methods. Moreover, the experimental validation shows the learned parameters obtain an accuracy that is similar to the off-line method in few iterations and having an important data reduction.

Acknowledgements. This research is supported by projects TIN2016-77836-C2-1-R, DPI2016-78957-, H2020-ICT-2014-1-644271, H2020-NMBP-TO-IND-2018-814426.

References

1. Algabli, S., Serratosa, F.: Embedding the node-to-node mappings to learn the graph edit distance parameters. Pattern Recogn. Lett. **112**, 353–360 (2018)
2. Caetano, T.S., McAuley, J.J., Cheng, L., Le, Q.V., Smola, A.J.: Learning graph matching. IEEE Trans. Pattern Anal. Mach. Intell. **31**(6), 1048–1058 (2009)
3. Conte, D., Foggia, P., Sansone, C., Vento, M.: Thirty years of graph matching in pattern recognition. Int. J. Pattern Recogn. Artif. Intell. **18**(03), 265–298 (2004)
4. Cortés, X., Serratosa, F.: Learning graph-matching edit-costs based on the optimality of the oracle's node correspondences. Pattern Recogn. Lett. **56**, 22–29 (2015)
5. Cortés, X., Serratosa, F.: Learning graph matching substitution weights based on the ground truth node correspondence. Int. J. Pattern Recogn. Artif. Intell. **30**(02), 1650005 (2016)

6. Gao, X., Xiao, B., Tao, D., Li, X.: A survey of graph edit distance. Pattern Anal. Appl. **13**(1), 113–129 (2010). https://doi.org/10.1007/s10044-008-0141-y
7. Leordeanu, M., Sukthankar, R., Hebert, M.: Unsupervised learning for graph matching. Int. J. Comput. Vision **96**(1), 28–45 (2012)
8. Lloyd, S.: Least squares quantization in PCM. IEEE Trans. Inf. Theor. **28**(2), 129–137 (2006). https://doi.org/10.1109/TIT.1982.1056489
9. Moreno-García, C.F., Cortés, X., Serratosa, F.: A graph repository for learning error-tolerant graph matching. In: Robles-Kelly, A., Loog, M., Biggio, B., Escolano, F., Wilson, R. (eds.) S+SSPR 2016. LNCS, vol. 10029, pp. 519–529. Springer, Cham (2016). https://doi.org/10.1007/978-3-319-49055-7_46
10. Neuhaus, M., Bunke, H.: Self-organizing maps for learning the edit costs in graph matching. IEEE Trans. Syst. Man. Cybern. Part B (Cybern.) **35**(3), 503–514 (2005)
11. Neuhaus, M., Bunke, H.: Automatic learning of cost functions for graph edit distance. Inf. Sci. **177**(1), 239–247 (2007)
12. Sanfeliu, A., Alquézar, R., Andrade, J., Climent, J., Serratosa, F., Vergés-Llahí, J.: Graph-based representations and techniques for image processing and image analysis. Pattern Recogn. **35**, 639–650 (2002)
13. Serratosa, F.: Fast computation of bipartite graph matching. Pattern Recogn. Lett. **45**, 244–250 (2014)
14. Serratosa, F.: Speeding up fast bipartite graph matching through a new cost matrix. Int. J. Pattern Recogn. Artif. Intell. **29**, 1550010 (2014). https://doi.org/10.1142/S021800141550010X
15. Serratosa, F.: A methodology to generate attributed graphs with a bounded graph edit distance for graph-matching testing. IJPRAI **32**(11), 1850038 (2018)
16. Serratosa, F.: Graph edit distance: restrictions to be a metric. Pattern Recogn. **90**, 250–256 (2019)
17. Serratosa, F., Cortés, X.: Interactive graph-matching using active query strategies. Pattern Recogn. **48**(4), 1364–1373 (2015)
18. Stauffer, M., Tschachtli, T., Fischer, A., Riesen, K.: A survey on applications of bipartite graph edit distance. In: Foggia, P., Liu, C.-L., Vento, M. (eds.) GbRPR 2017. LNCS, vol. 10310, pp. 242–252. Springer, Cham (2017). https://doi.org/10.1007/978-3-319-58961-9_22
19. Vento, M.: A one hour trip in the world of graphs, looking at the papers of the last ten years. In: Kropatsch, W.G., Artner, N.M., Haxhimusa, Y., Jiang, X. (eds.) GbRPR 2013. LNCS, vol. 7877, pp. 1–10. Springer, Heidelberg (2013). https://doi.org/10.1007/978-3-642-38221-5_1

Congratulations!
Dual Graphs Are Now Orientated!

Darshan Batavia[1]([✉]), Walter G. Kropatsch[1], Rocio M. Casablanca[2],
and Rocio Gonzalez-Diaz[2]

[1] Pattern Recognition and Image Processing Group 193/03, TU Wien,
Vienna, Austria
{darshan,krw}@prip.tuwien.ac.at
[2] Applied Math I, University of Seville, Seville, Spain
{rociomc,rogodi}@us.es

Abstract. A digital image can be perceived as a 2.5D surface consisting
of pixel coordinates and the intensity of pixel as height of the point in the
surface. Such surfaces can be efficiently represented by the pair of dual
plane graphs: neighborhood (primal) graph and its dual. By defining orientation of edges in the primal graph and use of Local Binary Patters
(LBPs), we can categorize the vertices corresponding to the pixel into
critical (maximum, minimum, saddle) or slope points. Basic operation of
contraction and removal of edges in primal graph result in configuration
of graphs with different combinations of critical and non-critical points.
The faces of graph resemble a slope region after restoration of the continuous surface by successive monotone cubic interpolation. In this paper,
we define orientation of edges in the dual graph such that it remains
consistent with the primal graph. Further we deliver the necessary and
sufficient conditions for merging of two adjacent slope regions.

1 Introduction

Configuration of such critical points and slope lines of a surface in term's of earth
topography were discussed in [2,11]. Lee in [10] investigated the configurations
of critical points of a Morse function of two variables with it's graphical representation. Moving a century ahead, in [3], authors used the neighborhood graph
and explained the use of Local Binary Patterns (LBPs) in predicting the critical
(maxima, minima and saddle) points and the slope points in digital images. By
performing contraction and removal operations on edges, they formed a stack of
graphs called graph pyramid. Cerman *et al.* [4] provide a practical application of
multi-resolution image segmentation using graph pyramids. Similar approaches
are used by Wei in [12] where a hierarchical structure similar to graph pyramid were constructed by using superpixels. The literature except [3] does not
consider the topological aspect of the surfaces which are covered by the papers
mentioned in the following paragraph.

In [6,7] Edelsbrunner et al. propose an algorithm of constructing a hierarchy
of increasingly coarse Morse-Smale complexes to decompose a piecewise linear

© Springer Nature Switzerland AG 2019
D. Conte et al. (Eds.): GbRPR 2019, LNCS 11510, pp. 131–140, 2019.
https://doi.org/10.1007/978-3-030-20081-7_13

2D-manifold with all its critical points being distinct. In our previous research work [8,9], we further generalize this concept beyond Morse-Smale complex and present a new hierarchy of increasingly coarse complexes decomposing 2D continuous surfaces denoted as slope complexes.

We also discussed properties of monotonic paths and provided a formula to count the number of slope regions at the top level of the pyramid with all its critical points being distinct. Our main aim was to preserve the critical points of the surface at the top level of the pyramid and to connect them with the minimum number of slope regions calculated using Euler's formula [5, Theorem 4.2.7].

In the past, the graph pyramids [3] are build on the top of the pair of dual graphs. The contraction of edges generates self-loops and multiple edges. The empty self-loops and multiple edges can be removed and simplified. This simplifications are controlled by the dual graphs. These dual graphs were not oriented and also were not used to capture the non topological properties of the graph. The main contribution of this paper is to provide orientation to the dual graph. A first step in this direction was done in [9] but here we provide it's interpretation in the image context and give properties of the oriented dual graphs to provide a reduction technique to meet our aims.

The paper is organized as follows: We start with the basic definitions in the field of images from the topological point of view. In Sect. 3, we define a concept to orient the dual graphs and the consistency of LBP categories. We also introduce a technique to orient the monotonic paths which contains level curves in the primal graphs. Section 4 provides necessary and sufficient conditions for both primal and dual graph to merge two slope regions. In Sect. 5, we show some experimental results for multi-resolution image segmentation. We end the paper with a note of what is attained from the paper and the possible extensions.

2   Orienting the Primal Graph

This section provides necessary definitions that form the basis of the further document. A discrete 2D image P where the intensity of a pixel p denoted by $g(p)$, can be represented as a neighborhood graph $G = (V, E)$ also referred to as primal graph. Every pixel p in the image P corresponds to a vertex $v \in V$ with gray value (g-value) $g(v) := g(p)$. Vertex v is connected to it's four adjacent vertices by edges $e \in E$.

The dual of primal graph G is denoted by $\overline{G} = (\overline{V}, \overline{E})$, where $\overline{V}$ being the vertex set of $\overline{G}$ which is associated with the faces of G, while $\overline{E}$ is the edge set of $\overline{G}$ which corresponds to the borders separating the faces of G. In other words, there is an edge $\overline{e}$ in the dual graph $\overline{G}$ for every edge e in primal graph G as mentioned in [5, Section 4.6]. There is a one-to-one correspondence between the edges of G and $\overline{G}$ so as the faces of G and vertices of $\overline{G}$. By performing the contraction and removal operations successively on the graph G, we obtain a stack (pyramid) of successively reduced plane[1] graphs $(G_k, \overline{G_k}), k \in [1, 2, \ldots, n]$.

---

[1] There is a topological and a combinatorial isomorphism between G and $\overline{G}$ and it is a unique pair of graphs embedded in a surface [5, pp. 70-80].

The base level of this pyramid is denoted as $G = G_0$. A Contraction operation in $G_k, 0 \leq k \leq n - 1$ corresponds to removal in $\overline{G_k}$ and a removal operation in G_k (merging of two faces) corresponds to contraction of two vertices in $\overline{G_k}$.

Definition 1. *The **orientation of an edge** $(v, w) \in E$ **in the primal graph** $G = (V, E)$ is directed from vertex $v \in V$ to vertex $w \in V$ iff $g(v) > g(w)$, all the other edges are not oriented.*

Orientation of edges can now be used to categorize the vertex $v \in V$ into critical (maximum, minimum, saddle) or slope point.

Definition 2. *A vertex $v \in V$ is a local **local maximum** $\oplus$ if all the edges incident to v are oriented outwards.*

Definition 3. *A vertex $v \in V$ is a **local minimum** $\ominus$ if all the edges incident to v are oriented inwards.*

Definition 4. *A vertex $v \in V$ is a **saddle point** $\otimes$ if there are more than two changes in the orientation of edges when traversed circularly (clockwise or counter-clockwise direction).*

Definition 5. *A vertex $v \in V$ is a **slope point** if there are exactly two changes in the orientation of edges when traversed circularly (clockwise or counter-clockwise direction).*

Categorizing a vertex using orientation of edges incident to it is equivalent to that of LBP code. The LBP value of an outward oriented edges are encoded as 1 and inward orientated edges are encoded as 0. The LBP code of a vertex is formed by concatenating LBP values of the incident edges in clockwise or counter-clock wise direction. The LBP code of a maximum will consist of 1 only while the LBP code of a minimum will consist of 0 only. The LBP code of slope points will have exactly 2 bit switches and saddles will have more than 2 bit switches. By use of orientated edges, we avoid the calculation of derivatives and eigen-values of the Hessian matrix to categorize a vertex.

Definition 6. *A **path** $\pi(v_1, v_r) = (V_\pi, E_\pi)$ is a non empty sub-graph of $G = (V, E)$, where $V_\pi = \{v_1, \ldots, v_r\} \subset V$ and $E_\pi = \{(v_1, v_2), \ldots, (v_{r-1}, v_r)\} \subset E$.*
*A path $\pi(v_1, v_r)$ is a **monotonic path** if all the oriented edges $(v_i, v_{i+1}), i \in [1, r - 1]$ have a same orientation.*

Remark 1. All the oriented edges on a monotonic path have the same orientation consequently defining the orientation of a monotonic path. Observe that if $g(v_i) = g(v_{i+1}), \forall i \in [1, r - 1]$ is called a *level curve* and it is a special case of monotonic paths.

A monotonic path $\pi(v_1, v_r)$ can be further extended by adding an edge oriented in the same direction as the direction of monotonic path $\pi(v_1, v_r)$. A monotonic path which cannot be further extended is called a **maximal monotonic path**. The end points of a maximal monotonic path will always be a local maximum and a local minimum.

2.1 Contracting Plateaus

A connected sub-graph having the same g-value for all the vertices is referred
to as a **plateau region** where every pair of vertices $v, w \in V$ of the sub-graph
satisfies $g(v) = g(w)$. The LBP encoding (see next subsection) is performed after
contraction of all the edges in the plateau regions where each plateau region
collapses either to a single vertex in the best case or a set of self-loops attached
to a vertex, surrounding every hole in the plateau region [8].

Nevertheless, the vertices on the boundary of the image need to be treated
differently. First, the case where the plateau region is connected to the boundary
as shown by the shaded region in the left images of Fig. 1(a), (b), needs to be
treated specially. To preserve the topology, we first perform contraction on the
vertices corresponding to the pixel on the boundary of the image.

As a result we get the vertices through which the border is connected to the
plateau region, and then collapse the remaining part of the plateau region into
a level curve as shown in the right images of Fig. 1(a), (b). Simultaneously, it
also explains the reason to perform the operations on border independently and
prior to the edges encapsulated by the border.

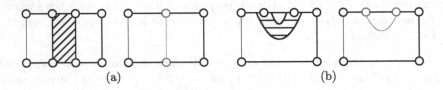

(a)   (b)

Fig. 1. Plateau regions connected to the boundary.

3 Orientation of Dual Graphs

In this section, we introduce a method to orient the edges of the dual graph corresponding to the oriented edges of the primal graph and explore it's properties
to help merging of faces in the primal graph. Maintaining generality, $\overline{e}$ in $\overline{G_h}$ is
normal to the tangent at any point on the edge e in G_h.

Definition 7 (Orientation of edges in dual graph). *The orientation of an*
edge $\overline{e}$ in $\overline{G}$ is from $\overline{v_r}$ to $\overline{v_l}$, where $\overline{v_r}$ and $\overline{v_l}$ are the vertices in $\overline{G}$ corresponding
to the faces on the right and the left side of the respective edge e in G while
walking in the direction of orientation of e.

Remark 2. The condition where the LBP code of $\overline{v}$ in $\overline{G}$ consists of only 0^*
or only 1^* can never exist as it corresponds to a cyclic sub-graph in G_k and
it contradicts the orientation of edges initially defined for the primal graph.
Nevertheless, directed cycles in the dual graph can appear and they surround
extrema in the primal.

Presence of level curves in G_k may complicate the orientation of the corresponding edge in $\overline{G_k}$ which we address later in this document.

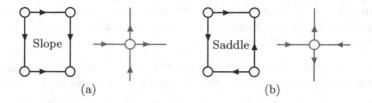

Fig. 2. Orientated dual graph for slope and saddle regions in primal graph.

Definition 8 (Slope region). *A face in G_k is a slope region iff the corresponding vertex $\overline{v}$ in $\overline{G_k}$ is a slope point.*

Imagine the primal graph G_0 as a terrain with height corresponding to the g-value of the vertex, then a face in G_0 will be a slope region or a saddle region[2] in the terrain. Definition 7 remains consistent to this concept and gives a matching LBP category of a vertex in dual as shown in Fig. 2.

Lemma 1 ([9]). *The boundary of a slope region S in the primal graph G is either composed of exactly two monotonic paths connecting two extrema or it is a level curve.*

Lemma 2 [1, Lemma 1]. *After contracting plateaus and adding dummy saddle points inside non-well composed configurations, all the vertices in the dual graph are slope points.*

4   Merging of Two Slope Regions

This section starts with basic requirements for merging of two slope regions. Then we provide the prototype of two adjacent slope regions sharing a common boundary. Finally we enumerate all the possible configurations of two slope regions deduced from the prototype and provide the constructive statements which are the necessary conditions for merging of two slope regions.

Two slope regions sharing a common boundary can be merged together by removal of the common boundary which results to form a **merged slope region**. We do not constrain the number of vertices and edges on the boundary of the slope regions. Hence the boundaries of the slope regions are referred as paths which may contain more than one vertices and edges. From Lemma 1 and properties of monotonic paths, the boundary of a slope region consist of exactly one level curve or two monotonic paths connecting one maximum and one minimum 'with respect to the slope region'. Note that by the usage of term 'with respect to the slope region', we constrain the connections of the vertex with the interior and the boundary of the slope region. We do not consider the connections of the vertex with the remaining graph where it can be categorized into a different LBP category.

---

[2] Region with a non well-composed configuration which requires insertion of a saddle point [3].

Merging of two slope regions may not necessarily result in a slope region. The merging of two slope regions can be done by checking whether the resulting merged region is a slope region or not.

Remark 3. Two adjacent slope regions in the primal graph can be merged iff the dual vertex corresponding to the merged slope region would be a slope point.

According to Lemma 1, the boundary of a slope region should be composed of two separate monotonic paths connecting the maximum and minimum of the slope region. The vertices at the end point of the common boundary between the two slope regions are already part of two monotonic paths, one from each slope region. To be a part of a monotonic path, these two vertices should either be an extremum or a slope point with respect to the monotonic path. If this condition is violated, it will contradict to the orientation of the monotonic path on the boundary of the resulting slope region. Now the proof of this boils down to demonstrate that the dual vertex corresponding to a face (in the primal graph) surrounded by exactly two monotonic paths is a slope point. Considering the circular permutation of the LBP codes of above mentioned dual vertex, we will have exactly two switches which as per Definition 5 is a slope point.

Let us consider sub-graphs of two slope regions S_1 and S_2 with their extrema $\oplus_1$, $\ominus_1$, $\oplus_2$ and $\ominus_2$ respectively. While formulating rules for merging two slope regions, the position of an extrema on the boundaries and it's connection with the common boundary are the main features to be considered. Besides extrema, the boundaries are composed of slope points *with respect to the slope region*. In this way we provide conditions which are independent of the number of edges and vertices on the boundary of the slope regions S_1 and S_2.

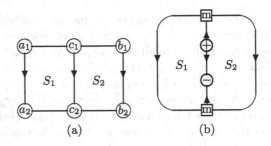

Fig. 3. Prototype for merging of two slope regions.

Figure 3a shows the prototype of two slope regions with a common boundary $\pi(c_1, c_2)$ between them. The paths connecting the vertices in the prototype may consist of any number of vertices and edges completing the respective connection. The orientation of paths $\pi(a_1, a_2)$, $\pi(b_1, b_2)$ and $\pi(c_1, c_2)$ is predefined by assuming that $\oplus_1 \in \{a_1, c_1\}$, $\oplus_2 \in \{b_1, c_1\}$, $\ominus_1 \in \{a_2, c_2\}$ and $\ominus_2 \in \{b_2, c_2\}$ positions. The theory remains the same if the positions of $\oplus_1$ and $\oplus_2$ are interchanged with $\ominus_1$ and $\ominus_2$ thereby reversing the orientation (flipping) of respective edges[3].

---

[3] This configuration can be achieved by switching positions of $\oplus_i$ and $\ominus_i$ in the previously mentioned configuration.

Moreover, all the combinations where the orientation of path $\pi(a_1, a_2)$ and $\pi(b_1, b_2)$ are opposite, for example $\oplus_1$ at a_1 and $\oplus_2$ at b_2 will not be valid because it contradicts the orientation of path $\pi(c_1, c_2)$ unless it is a level curve (which is taken into account in Case 5). By putting restrictions on positions, we reduce the number of possible combinations to $2^4 = 16$ configurations, which can be further reduced (by interchanging S_1 and S_2) for investigative purpose and still keeping the rules general.

The removal of a path $\pi(c_1, c_2)$ with n edges consists of following two steps:

1. Contract $(n-1)$ edges as a result of which the path will be composed of only 1 edge connecting the end points.
2. Perform the removal operation on the remaining edge.

Following are enumerations in which the two slope regions S_1 and S_2 can be merged such that the resulting slope region will still obey Definition 8 and preconditions of Lemma 1. Preserving the condition, the number of different configurations which can be generated by interchanging labels S_1 and S_2 are mentioned in the brackets after each condition.

Case 1: We start with the most simple case of Fig. 4a, where S_1 and S_2 share the same extrema on the common boundary, i.e. $\oplus_1 = \oplus_2$ at c_1 and $\ominus_1 = \ominus_2$ at c_2. In this case, remaining paths are directed from c_1 to c_2, one through a_1 and a_2, and the other through b_1 and b_2 [2 combinations].

Case 2: If the common boundary is composed of one extremum from S_1 and one extremum from S_2 as shown in Fig. 4b [2 combinations].

Case 3: If both of the end points of the common boundary contains extrema from a single slope region irrespective of the position of extrema from other slope region as shown in Fig. 5a. For example: $\oplus_1$ at a_1 and $\ominus_1$ at a_2 respectively while $\oplus_2$ and $\ominus_2$ contribute to the end points of the common boundary [6 combinations].

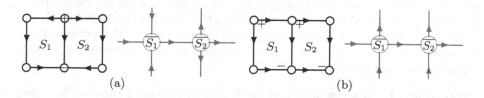

(a)   (b)

Fig. 4. Slope regions contributing at-least one extremum to the common boundary.

Case 4: None of the extrema contribute to the common boundary, but both extrema of one slope region are connected to the common boundary through a level curve as shown in Fig. 5b. The blue vectors besides the graph shows the orientation of the monotonic path which consist of level curves [1 combination].

Case 5: If one extremum of a slope region contributes to the end point of the common boundary and the other is connected to the common boundary through a level curve irrespective of the position of extrema of the other slope region. For example refer to Fig. 5c, d. The dashed line in the figure refers to the level curve and the blue vectors besides the graphs shows the orientation of the monotonic path [6 combinations].

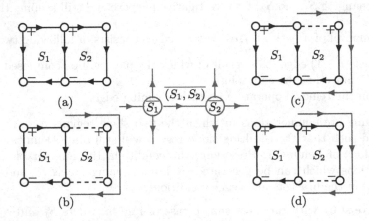

Fig. 5. At-least one extremum of a slope region is connected to the common boundary through a level curve.

In all the above cases, the common boundary path $\pi(c_1, c_2)$ has orientation in a single direction. We also take into consideration the cases where the common boundary have orientations of edges in different direction like shown in Fig. 3b, where the vertex m is a slope point with respect to the slope regions. In its local neighborhood, m can be a slope point or a saddle point. Note: This case is counted under the condition where both the slope regions share common extrema and hence the slope regions can be merged. There may be cases where only one of the two common slope points m is present and the monotonic paths are directly connected to an extremum. Removal of path $\pi(\oplus, \ominus)$ in Fig. 3b will result in pending edges connecting m and the extrema in the primal graph. Corresponding to this pending edge, there will a self-loop surrounding a single vertex in the dual graph.

Proposition 1. *Any two slope regions S_1 and S_2 sharing a common boundary can be merged together if they follow one of the following condition:*

1. *The common boundary is composed of at least one extremum of S_2 and another extremum is either on the common boundary or is connected to the common boundary through a level curve.*
2. *The common boundary is composed of one extremum from each slope region.*

For all the cases in Figs. 4b and 5, the edges incident to the dual vertex $\overline{S_1}$ have the same orientation as the dual vertex $\overline{S_2}$ when traversed in a circular (clockwise or counter clockwise) order. Similar observations can be made in the corresponding primal graphs.

In other words, both of the vertices in the dual exhibit the same LBP code after eliminating redundancy of bits when traversed in same direction. The vertex in Fig. 6 is a slope point formed as a result of the contraction operation on the dual graph in Fig. 5. Similarly we get a slope point after contracting the dual edge connecting the two dual vertices in Fig. 4b.

Fig. 6. Result after contraction of edge $\overline{(S_1, S_2)}$ in Fig. 5

Few examples can be viewed in Fig. 7 where the extrema are connected to the common boundary without a level curve or the orientation of the level curve is reversed. We observe that the vertices in the dual graph have different LBP codes when traversed in the same direction.

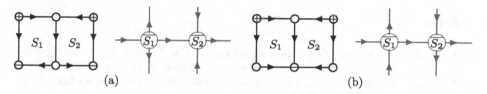

Fig. 7. Configurations of slope regions that cannot be merged and their oriented dual graphs.

Proposition 2. *The contraction of the dual edge $\overline{e}$ connecting two slope points $\overline{v_1}$ and $\overline{v_2}$ in dual graph $\overline{G}$ will result in a single slope point iff the edges incident on the vertices $\overline{v_1}$ and $\overline{v_2}$ exhibit the same orientation (the same LBP code) between the bit switches when traversed in same order and direction.*

Proposition 2 provides sufficient conditions for contraction of edge $\overline{e}$ in $\overline{G}$ connecting two slope points such that the resulting dual vertex is also a slope point. This contraction is equivalent to the removal of the corresponding edge e in G such that the resulting region is a slope region.

4.1   Orientation of Level Curves Shared by Multiple Monotonic Paths

Technically level curves follow the surface along the same height. Hence they are not oriented. The corresponding (level) paths can be concatenated with adjacent monotonic paths if all the involved monotonic paths have the same orientation. They then form a combined monotonic path in which the level paths inherit their orientation from the orientation of the combined monotonic path. This may lead to inconsistencies if the same level path or a sub-sequence of it is simultaneously concatenated with a monotonic path of the opposite orientation. In such cases preference will be given to the orientation of the level path which is involved in the merging of slopes. The priority of merging slopes may depend on higher objectives like making slope regions of the global extrema as large as possible.

5 Conclusion

In this paper, we represent a surface by a pair of dual plane graphs and provide a novel solution of collapsing the plateau region which basically is the collection of degenerated surface points. Then we categorize the vertices into maximum, minimum, slope or saddle depending on the orientation of the incident edges and avoiding the calculations of differentiation. We also define the orientation of the edges in the dual graph and show that the LBP category of the dual vertex is consistent with the corresponding face in the primal graph. Then we give the necessary and sufficient conditions for merging of two adjacent faces in the primal graph of a well composed sampled surface such that the merged face is a slope region. Finally we offer the sufficient conditions for the resulting dual vertex to be a slope point produced after contracting the respective dual edge.

References

1. Batavia, D., Kropatsch, W.G., Gonzalez-Diaz, R., Casblanca, R.M.: Counting slope regions in surface graphs. In: Computer Vision Winter Workshop (2019)
2. Cayley, A.: Xl. on contour and slope lines. London, Edinburgh, Dublin Philos. Mag. J. Sci. **18**(120), 264–268 (1859)
3. Cerman, M., Gonzalez-Diaz, R., Kropatsch, W.: LBP and irregular graph pyramids. In: Azzopardi, G., Petkov, N. (eds.) CAIP 2015. LNCS, vol. 9257, pp. 687–699. Springer, Cham (2015). https://doi.org/10.1007/978-3-319-23117-4_59
4. Cerman, M., Janusch, I., Gonzalez-Diaz, R., Kropatsch, W.G.: Topology-based image segmentation using LBP pyramids. Mach. Vis. Appl. **27**(8), 1161–1174 (2016)
5. Diestel, R.: Graph Theory: Graduate Texts in Mathematics (1997)
6. Edelsbrunner, H., Harer, J., Natarajan, V., Pascucci, V.: Morse-smale complexes for piecewise linear 3-manifolds. In: Proceedings of the Nineteenth Annual Symposium on Computational Geometry, pp. 361–370. ACM (2003)
7. Edelsbrunner, H., Harer, J., Zomorodian, A.: Hierarchical morse-smale complexes for piecewise linear 2-manifolds. Discrete Comput. Geom. **30**(1), 87–107 (2003)
8. Kropatsch, W.G., Casablanca, R.M., Batavia, D., Gonzalez-Diaz, R.: Computing and reducing slope complexes. In: Marfil, R., Calderón, M., Díaz del Río, F., Real, P., Bandera, A. (eds.) CTIC 2019. LNCS, vol. 11382, pp. 12–25. Springer, Cham (2019). https://doi.org/10.1007/978-3-030-10828-1_2
9. Kropatsch W.G., Casablanca R.M., Batavia D., Gonzalez-Diaz, R.: On the space between critical points. In: Couprie, M., Cousty, J., Kenmochi, Y., Mustafa, N. (eds.) Discrete Geometry for Computer Imagery. DGCI 2019. LNCS, vol. 11414, pp. 115–126. Springer, Cham (2019). https://doi.org/10.1007/978-3-030-14085-4_10
10. Lee, R.N.: Two-dimensional critical point configuration graphs. IEEE Trans. Pattern Anal. Mach. Intell. **4**, 442–450 (1984)
11. Maxwell, J.C.: L. on hills and dales: to the editors of the philosophical magazine and journal. London, Edinburgh, Dublin Philos. Mag. J. Sci. **40**(269), 421–427 (1870)
12. Wei, X., Yang, Q., Gong, Y., Ahuja, N., Yang, M.H.: Superpixel hierarchy. IEEE Trans. Image Process. **27**(10), 4838–4849 (2018)

A Parallel Algorithm for Subgraph Isomorphism

Vincenzo Carletti$^{(\boxtimes)}$, Pasquale Foggia, Pierluigi Ritrovato, Mario Vento, and Vincenzo Vigilante

Department of Information Engineering, Electrical Engineering and Applied Mathematics, University of Salerno, Fisciano, Italy
{vcarletti,pfoggia,pritrovato,mvento,vvigilante}@unisa.it
http://mivia.unisa.it

Abstract. In different application fields, such as biology, databases, social networks and so on, graphs are a widely adopted structure to represent the data. In these fields, a relevant problem is the detection and the localization of structural patterns within very large graphs; such a problem, formalized as subgraph isomorphism, has been proven to be NP-Complete in the general case. Moreover, the continuously growing size of the graphs to face, actually of hundred thousands of nodes, is making the problem even more challenging also for the most efficient algorithms in the state of the art, requiring days or weeks of computational time. This huge amount of time is also consequence of the fact that most of the algorithms do not exploit any kind of parallelism, even if the problem is suitable to be solved adopting parallel approaches. In this paper we present a new parallel algorithm for subgraph isomorphism, namely VF3P, based on a redesign of the well known algorithm VF3. The effectiveness of VF3P has been experimentally proven on a publicly available dataset of very large graphs, confirming that the algorithm is able to efficiently scale w.r.t. the number of used CPUs without affecting the memory usage.

Keywords: Exact graph matching · Subgraph isomorphism · Parallel algorithms · VF3

1 Introduction

Graphs are discrete mathematical structures representing objects in terms of their parts and the relationships among these parts, using abstractions called nodes and edges respectively. Such a representation is much more expressive than the vector-based one, but it requires more complex algorithms, and thus a higher computational effort, also to perform simple operations like evaluating the similarity between two objects. Nevertheless, there are several cases where graphs are preferred to vectors because the latter are ineffective to model the complexity of the objects especially when these are composed by parts suitably

© Springer Nature Switzerland AG 2019
D. Conte et al. (Eds.): GbRPR 2019, LNCS 11510, pp. 141–151, 2019.
https://doi.org/10.1007/978-3-030-20081-7_14

interconnected each other and the application at hand exploits this relevant structural information [16, 18, 26].

Nowadays, the field of social networks [15, 27], databases, semantic web and biology require to use bigger and bigger structural information, typically represented in term of graphs [3, 21]. Among them, the latter is undoubtedly the most promising and challenging area [5, 9, 13, 14], where many biological entities are naturally represented as graphs; moreover, the quantity of data generated every year and needed to be analyzed, is enormous. Noteworthy examples are molecular and protein structures, interaction networks and more recently the genome, that for many years has been represented as a string of bases [4, 23].

In this context graph matching algorithms play an important role because they allow to perform the basic operations required to apply pattern recognition methods, such as the computation of the similarity (i.e. the distance) or the search for a structural pattern.

It is important to say that all the previously cited fields provide, year by year, new challenges to graph matching algorithms, due to the continuously growing size of the graphs they require to deal on. It it important to note that, actually, even a graph of thousand nodes is considered small in many cases, for instance when working with the genome of an individual that is composed of billions of bases. Therefore, even a frequent operation like searching for a pattern structure inside a graph, namely the subgraph isomorphism problem, becomes very time expensive also for the most efficient algorithms in the state of the art.

The current graph matching algorithms have been generally designed according to a sequential computational paradigm, even if many operations required by them can be potentially done in parallel; indeed, analyzing the literature it is possible to find extremely efficient algorithms, such as VF3 [8, 11, 12, 14] and RI [5], able to work with graphs of thousands of nodes using a very limited quantity of memory and CPU, but requiring a large amount of time when the size and the density of the graphs increase (e.g. weeks of computational time).

Another important evidence of the need for parallel graph matching algorithms, is the growing interest in parallelizing the computation of the graph edit distance recently arisen in the scientific community [1, 2, 6, 24].

Realizing efficient parallel graph matching algorithms is not as easy as it would appear. Indeed, starting from a sequential algorithm and making in parallel some of its steps and procedures is, in most of the case, useless. The big challenge is to design the algorithm so as to take full advantage form a specific parallel architecture, such as multicore systems using CPU, GPU, clusters and so on. Distributed subgraph isomorphism methods to deal with very large graphs have been proposed in [7, 28], but there still remains the problem of reducing the high communication cost among the nodes of the cluster. Concerning the GPUs, some papers [20, 29] have proposed interesting performance analysis on graph matching algorithms for GPU highlighting the bottlenecks and the reasons why graph algorithms are not able to exploit the architecture CPU-GPU at the best. A parallel approach on multicore CPUs has been recently proposed by McCreesh et al. [22]. The authors have proposed a simple parallel constrain programming

approach based on LAD [25]; they have focus the attention on specific paralleliz-
able steps and have presented an analysis limited to the execution time, without
discussing efficiency and speedup.

In this paper we propose a parallel algorithm for multicore CPUs obtained
from VF3-Light [11] by realizing a state-level parallelization. The effectiveness
of our proposal have been proved by analyzing the memory requirements, the
speed-up and the efficiency with respect to the original sequential algorithm.

2   VF3-Light: The Sequential Algorithm

In this section, we briefly present some fundamental concepts on graph matching and VF3-Light required to understand the design choices discussed in the
successive sections, the reader who is interested in deepening the algorithm is
referred to [8,11].

2.1   Graphs and Graph Matching

A graph is an ordered pair $G = (V, E)$ where V and $E \subset V \times V$ are the set of
nodes and edges respectively. Given a node $u \in V$, the set of its *successors* (the
nodes connected to u by outgoing edges) is denoted as $\mathcal{S}(u)$, while the set of
its *predecessors* (the nodes connected by incoming edges) is denoted as $\mathcal{P}(u)$. In
a more general definition, graphs can carry also *attributes* or *labels* attached to
their nodes and edges. Hence, two additional sets are considered: the set of node
labels L_v and the set of edge labels L_e; two labeling functions, $\lambda_v : V \to L_v$ and
$\lambda_e : E \to L_e$, are used to associate each node or edge to the corresponding label.

Considering two graphs, namely $G_1 = (V_1, E_1)$ and $G_2 = (V_2, E_2)$, graph
matching is the problem of finding a function $M : V_1 \to V_2$, namely the *mapping
function*, satisfying some structural constraints. In the case of subgraph isomorphism [16], the constraints are that M is injective and *structure preserving*,
i.e. the nodes put in correspondence must have the same structure considering
both the presence and the absence of edges. It is important to note that the mapping function is not unique, but the problem can have several distinct solutions
where the nodes in V_1 are mapped to different subsets of nodes in V_2. In general,
we are not interested in finding the first solution only, but all the possible ones.

2.2   VF3-Light

VF3-Light is the most recent successor of the well-know algorithm for subgraph
isomorphism VF2 [17]; it has been proposed in [11] as a lightened version of
VF3 [8] where some of the heuristics, required to deal with large and dense
graphs, have been relaxed to reduce the overall computational time when facing
small or sparse graphs. All the algorithms, designed from VF2, share the same
structure and use a depth-first approach search (DFS) over a tree-structured
search space of states. Each *state* s_n represents a partial mapping $M(s_n)$ between

the nodes in V_1 and those in V_2. Two additional sets $M_1(s_n) \subseteq V_1$ and $M_2(s_n) \subset V_2$ are used to represent the nodes of V_1 and V_2, respectively, that are in $M(s_n)$.

That said, the algorithm starts from a *root state* s_0 where the mapping $M(s_0)$ is empty and proceed until it reaches a *leaf state* s_l whose mapping $M(s_l)$ is *complete*, i.e. it involves all the nodes in V_1. The exploration from the root to a leaf proceed by extending the nodes involved in the mapping M; at each state s_n a new one is generate by adding to the mapping $M(s_n)$, a new couple of nodes $u_n \in V_1$, $v_n \in V_2$ that are not yet in $M(s_n)$. Only some of the couples are feasible to generate a state that is *consistent* with the constrains of the subgraph isomorphism and so are used to generate a new state. Finally, the leaves are also consistent, namely the *goal states*, represent the solutions of the problem.

In order to avoid an expensive bind search over the whole search space, VF3-Light uses the *feasibility rule* (Eq. 1) to explore the sub-space composed of only *consistent* states.

$$\texttt{IsFeasible}(s_n, u_n, v_n) = F_s(s_n, u_n, v_n) \wedge F_c(s_n, u_n, v_n) \tag{1}$$

The function F_s verifies the *semantic consistency* of node and edge labels (or attributes), while F_c verifies the *structural consistency*. Such a rule aims at ensuring that the addition of a couple (u_n, v_n) to a consistent state s_n will not produce an inconsistent state.

3   Parallel Algorithms

Designing a parallel algorithm starting from a sequential one is not an immediate task, but it requires to analyze different aspects. A widely adopted methodological approach, proposed by Ian Foster in [19], organizes the design of a parallel algorithm in four steps: Partitioning, Communication, Agglomeration and Mapping. The first is the arrangement of the data into discrete chunk of work that can be distributed to multiple tasks. Two basic ways to perform this process are: *domain decomposition*, aimed at decomposing the data into many small partitions to which parallel computation may be applied, and *functional decomposition* where the problem is decomposed in terms of operations that can be performed simultaneously. Once the decomposition have been defined, it is necessary to define how the task communicate each other; for instance if they work in a coordinated way or asynchronously. Then the agglomeration step aims at reducing the number of tasks generated during the partitioning in order to reduce communication costs, that is the most influencing factor in parallel algorithms performance. Finally, in the mapping step we define where each task is to run in order to minimize the execution time. For instance, we can map on the same processors tasks that need a strong communication while on different processors tasks able to run concurrently.

The exploration of the search space is a problem suitable to effectively exploit the data parallelism, thus, according to the partitioning strategies proposed by Foster, we have designed a parallel algorithm, namely $VF3P$, based on a domain decomposition where each task is responsible to explore a single state. It is worth to point out that we have considered each task as performed by a single thread, therefore we will use equivalently the terms thread and task. The communication is realized through a *global state stack*; a task extracts the state to explore from the stack, then puts the feasible states generated in it. Each task stops when no further states have to be explored, such a condition is verified when the global stack is empty and all the other tasks have finished to explore their last extracted state, thus no other state is going to be put in the stack. Adopting such a strategy, the agglomeration is implicitly realized by the way the tasks communicate each other through the global structure. An outline of the procedure performed by each task in $VF3P$ is shown in Fig. 1. On the one hand, the used strategy allows to reach a high efficiency because all the threads work for most of the time; but, on the other hand, it requires a high level high level of synchronization among the threads that can affect the efficiency when the number of threads grows. To deal with this problem, we have designed a further optimization of $VF3P$, namely $VF3P_{LS}$, aimed at reducing the synchronization through the use of a side state stack used privately by the tasks. Each task has its *local state stack* where it puts the generated states to be successively explored. Therefore, until a task does not need to access the global stack it is able to work independently from the others. It is worth noting that using the local stack only does not guarantees that the workload is balanced among all the thread, this can cause a loss of efficiency due to the fact that some tasks are unoccupied. To avoid this problem, the local stack has a limited size, thus, when it is full the task is force to put the exceeding

```
1: function VF3P-Task(s, G₁, G₂, Sg, out Results)
2:     if IsEmpty(Sg) then
3:         return CheckActiveTasks()
4:     s := PullFromStack(Sg)
5:     if IsGoal(s) then
6:         append M(s) to Results
7:     else
8:         for (uₙ,vₙ) ∈ NextCandidates(s, G₁, G₂)
9:             if IsFeasible(s, uₙ, vₙ) then
10:                 sₙ := ExtendState(s, uₙ, vₙ)
11:                 PushInStack(sₙ, Sg)
12:         NotifyEndOfExploration()
13:     return True
14: end
```

Fig. 1. Outline of $VF3P$ task procedure. Each task pull the next state to process from the global stack S_g. If the state is not a goal one, the task explores all of its descendant and put in S_g only those are feasibile. Since the condition **IsEmpty**(S_g) is not sufficient to guarantees that no more states have to be explored, each task notifies the start of the exploration when it pulls a state from the stack and uses the procedure **NotifyEndOfExploration** to communicates to the others when it finished. When S_g is empty and no more tasks are involved in exploring a state, than all the tasks will stop working.

states in the global stack. It is worth to note that, if the size of local state stack is configured taking into account the maximum depth of the search space (the size of the pattern graph) and the density of the two graphs, each task will be able to explore the space, from the root to leaves, without picking states from the global stack. Moreover, considering how the DFS works, each task maintains in its local stack the states corresponding to the higher levels of the state space, while it tends to put in the global stacks the states belonging to the lower levels.

The two algorithms, $VF3P$ and $VF3P_{LS}$, differ in the procedures **Pull-FromStack** and **PushInStack** (see Fig. 1). Indeed, while the tasks of $VF3P$ work directly using the global stack, in $VF3P_{LS}$ each a task checks firstly if the local stack is empty (full) before accessing the global stack to pull (put) a state.

4   Experiments

The benchmark of parallel algorithms is not limited to time and memory requirements; indeed, two relevant performance measures (see Eq. 2) are *speed-up* (Sp) and *efficiency* (Ef). The first represents the improvement of the execution time evaluated as the ratio between the run time T_s of the most efficient sequential algorithm and T_p, the one of the parallel algorithm. The second characterizes how the parallel algorithm is efficient in exploiting the available hardware resources and is obtained dividing the speed-up by the number of CPUs.

$$Sp = \frac{T_s}{T_p} \qquad\qquad Ef = \frac{Sp}{\#CPU} \qquad (2)$$

The goal for a parallel algorithm is to reach a *linear speed-up*, where the value of the ratio is exactly the number of used CPUs. Very rarely, it is also possible to witness a *superlinear speed-up*, when the speed-up ratio is higher than the number of the CPUs. In general, a linear speed-up is very difficult to achieve because it requires that all the CPUs have always the same amount of workload and are able to execute their task independently or with few interactions. Unfortunately, not for all the problems it is possible to design algorithms exposing a linear speed-up. This is the case of graph algorithms, where this difficulty is confirmed by the fact that, until now, they have not been implemented effectively on modern GPU architectures, that are designed to exploit algorithms suitable to exhibit a linear speed-up such as low level image processing ones.

As previously introduced, computing the speed-up and the efficiency requires to choose a reference sequential algorithm, that is usually selected among the most efficient ones solving the problem under analysis. In our case, since the proposed parallel algorithms have been realized starting from VF3-Light, it is the most suited to this purpose, even because it has been proved to be one of the most efficient subgraph isomorphism algorithms. Therefore, in our experiments we have computed the aforementioned performance measures by executing $VF3P$ and $VF3P_{LS}$ with 2, 4 and 8 working threads respectively, in order to evaluate how the performance measures evolve when the number of thread grows.

Table 1. Speed-up of the parallel algorithms over different target graph size and number of CPU cores employed.

Dataset		Target size	Speed-up					
			$VF3P$			$VF3P_{LS}$		
			2 core	4 core	8 core	2 core	4 core	8 core
$\eta = 0.2$	Uniform	1000	0.76	0.75	0.70	0.99	0.84	0.76
		2000	1.59	2.34	3.42	1.55	2.63	3.74
		4000	1.56	2.96	5.16	1.69	3.14	5.61
		10000	1.79	3.44	6.36	1.82	3.53	6.63
	Non-Uniform	1000	0.70	0.75	0.61	1.06	0.76	0.57
		2000	1.54	2.16	2.98	1.51	2.33	3.11
		4000	1.70	2.98	4.77	1.72	3.12	4.97
		10000	1.77	3.37	6.11	1.80	3.47	6.41
$\eta = 0.3$	Uniform	1000	1.72	2.63	3.46	1.79	2.69	3.59
		2000	1.46	3.00	5.19	1.54	3.24	5.64
		4000	1.78	3.36	6.05	1.81	3.47	6.39
		6000	1.85	3.56	6.68	1.88	3.65	6.89
		8000	1.88	3.64	6.82	1.90	3.69	6.96
	Non-Uniform	1000	1.65	2.53	2.93	1.77	2.55	3.17
		2000	1.59	2.93	4.91	1.66	3.15	5.30
		4000	1.76	3.27	5.69	1.77	3.38	6.28
		8000	1.86	3.60	6.74	1.88	3.66	6.91

The experimental environment has been properly configured so as to collect unbiased measures and ensure that each thread run on the same core during all the execution time. The experiments have been performed on a Ubuntu 18.04 server where all the unnecessary services and the swap area have been deactivated. The server is equipped with two Intel(R) Xeon(R) CPU E5-2650 v2 and 256 Gb of Ram. Each Xeon E5-2650 has 8 physical core and three level of cache, in particular 256 Kb of L2 cache dedicated to the each single core and 20 Mb of L3 cache shared by all the cores laying on the same CPU. Hyperthreading has been deactivated to let the operating system to run one thread per physical core. One of the CPUs hosted the threads of the operating system and the experimental environment, while the other has been completely dedicated to run the working threads of the algorithms; thus, by setting the affinity we have ensured that each thread was executed on a dedicated core. In this way, we are able to properly measure how the algorithms improves speed-up and efficiency w.r.t. the number of cores by setting the wanted number of running threads.

The experiments have been performed over a subset of the MIVIA LDG, a standard dataset firstly used in [8,10,11] to benchmark VF3. The dataset is composed of very large and dense random Erdős and Rényi graphs, both labelled

Table 2. Efficiency of the parallel algorithms over different target graph size and number of CPU cores employed.

Dataset		Target size	Efficiency					
			$VF3P$			$VF3P_{LS}$		
			2 core	4 core	8 core	2 core	4 core	8 core
$\eta = 0.2$	Uniform	1000	0.38	0.18	0.08	0.49	0.21	0.09
		2000	0.79	0.58	0.42	0.77	0.65	0.47
		4000	0.78	0.74	0.64	0.84	0.78	0.70
		10000	0.89	0.86	0.79	0.91	0.88	0.83
	Non-Uniform	1000	0.35	0.18	0.07	0.53	0.19	0.07
		2000	0.77	0.54	0.37	0.75	0.58	0.38
		4000	0.85	0.74	0.59	0.86	0.78	0.62
		10000	0.88	0.84	0.76	0.90	0.86	0.80
$\eta = 0.3$	Uniform	1000	0.86	0.65	0.43	0.90	0.67	0.45
		2000	0.73	0.75	0.64	0.77	0.81	0.71
		4000	0.89	0.84	0.75	0.90	0.87	0.80
		8000	0.94	0.91	0.85	0.95	0.92	0.87
	Non-Uniform	1000	0.83	0.63	0.38	0.89	0.63	0.40
		2000	0.79	0.73	0.61	0.83	0.78	0.66
		4000	0.88	0.82	0.71	0.89	0.85	0.79
		8000	0.93	0.90	0.84	0.94	0.92	0.86

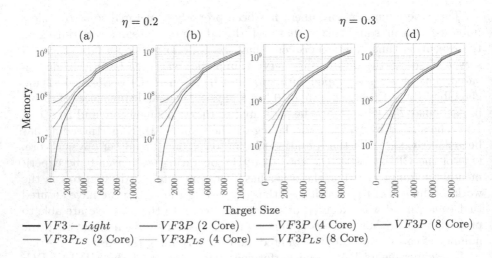

Fig. 2. Memory usage on $\eta = 0.2$ and $\eta = 0.3$ random graphs of both $VF3P$ and $VF3P_{LS}$, by varying the number of used cores.

and unlabelled, having densities (η) of 0.2, 0.3 and 0.4 respectively. The target graphs size ranges from 300 to 10, 000 nodes, while the size of the pattern graphs is 20% with respect to that of the corresponding target. As discussed in [10], although in the MIVIA LDG dataset there is only one solution for each pair of graphs, the computational effort required to search for the solution and explore the whole search space to confirm the absence of other solutions is very high. Therefore, due to its complexity such a dataset is very suitable to stress the algorithm in terms of CPU usage, so as to highlight possible loss of efficiency.

In Tables 1 and 2 we present the results of the experiments in terms of speed-up, efficiency on the considered datasets. Analyzing the speed-up it is possible to note that both the parallel algorithms are able to scale w.r.t. the number of cores. In particular, in the case of $\eta = 0.2$, the best speed-up is achieved by $VF3P_{LS}$ for graphs larger than 8, 000 node, and it is of 1.7, 3.5 and 6.6 when using 2, 4 and 8 cores respectively. On $\eta = 0.3$ graph, the best results is still obtained by $VF3P_{LS}$; it is worth to note that, in this case, the speed-up using 8 cores is about 7. The effectiveness of the proposed algorithms is also confirmed by the achieved efficiency, especially when the CPUs is more stressed, i.e. for graphs larger than 4, 000 nodes; as expected, both the algorithm are able to obtain values higher than 0.8 irrespective from the number of cores. Of course, using less cores the efficiency is higher due to the lower time lost in synchronization. Anyway, when the number of cores increases, the benefit of the local stack, adopted by $VF3P_{LS}$, in reducing the amount of synchronization time lost, is more evident both on the speed-up and on the efficiency.

Differently happens for the memory, there are not notable benefits in using the local stack because both the solutions requires the same amount of memory to manage the communication; indeed looking at the Fig. 2 the curves are completely overlapped. It is worth to note that, the higher is the number of cores the higher is the memory usage because of the higher is the number of states generated and buffered in the communication stacks; even if the growth in the usage of memory is very limited if compared with that of VF3-Light.

5   Conclusions

In this paper we have proposed $VF3P$, a parallel algorithm to solve subgraph isomorphism. The effectiveness of the proposed algorithm has been proved using very large and dense graphs considering three performance measures: the speed-up, the efficiency and the memory usage. On the base of the achieved results we have demonstrated that the proposed algorithm is very efficient and able to scale w.r.t. the number of used CPUs. Nevertheless, a deeper analysis can be performed to explore other aspects impacting the performance and further improvements to the efficiency can be achieved by adopting different communication schemas and agglomeration.

References

1. Abu-Aisheh, Z., et al.: Graph edit distance contest: results and future challenges. Pattern Recogn. Lett. **100**, 96–1103 (2017)
2. Abu-Aisheh, Z., Raveaux, R., Ramel, J.Y., Martineau, P.: A parallel graph edit distance algorithm. Expert Syst. Appl. **94**, 41–57 (2018)
3. Aittokallio, T., Schwikowski, B.: Graph-based methods for analysing networks in cell biology. Briefings Bioinform. **7**, 243–255 (2006)
4. Computational pan-genomics: status, promises and challenges. Oxford J. Brief. Bioinf. **19**, 118–135 (2016)
5. Bonnici, V., Giugno, R., Pulvirenti, A., Shasha, D., Ferro, A.: A subgraph isomorphism algorithm and its application to biochemical data. BMC Bioinform. **14**, 1–13 (2013)
6. Bougleux, S., Brun, L., Carletti, V., Foggia, P., Gazre, B., Vento, M.: Graph edit distance as a quadratic assignment problem. Pattern Recogn. Lett. **87**, 38–46 (2017)
7. Broecheler, M., Pugliese, A., Subrahmanian, V.S.: COSI: cloud oriented subgraph identification in massive social networks. In: 2010 International Conference on Advances in Social Networks Analysis and Mining (2010)
8. Carletti, V., Foggia, P., Saggese, A., Vento, M.: Challenging the time complexity of exact subgraph isomorphism for huge and dense graphs with VF3. IEEE Trans. Pattern Anal. Mach. Intell. **40**, 804–818 (2018)
9. Carletti, V., Foggia, P., Vento, M., Jiang, X.: Report on the first contest on graph matching algorithms for pattern search in biological databases. In: Liu, C.-L., Luo, B., Kropatsch, W.G., Cheng, J. (eds.) GbRPR 2015. LNCS, vol. 9069, pp. 178–187. Springer, Cham (2015). https://doi.org/10.1007/978-3-319-18224-7_18
10. Carletti, V., Foggia, P., Greco, A., Saggese, A., Vento, M.: Comparing performance of graph matching algorithms on huge graphs. Pattern Recogn. Lett. (2018)
11. Carletti, V., Foggia, P., Greco, A., Saggese, A., Vento, M.: The VF3-light subgraph isomorphism algorithm: when doing less is more effective. In: Bai, X., Hancock, E.R., Ho, T.K., Wilson, R.C., Biggio, B., Robles-Kelly, A. (eds.) S+SSPR 2018. LNCS, vol. 11004, pp. 315–325. Springer, Cham (2018). https://doi.org/10.1007/978-3-319-97785-0_30
12. Carletti, V., Foggia, P., Saggese, A., Vento, M.: Introducing VF3: a new algorithm for subgraph isomorphism. In: Foggia, P., Liu, C.-L., Vento, M. (eds.) GbRPR 2017. LNCS, vol. 10310, pp. 128–139. Springer, Cham (2017). https://doi.org/10.1007/978-3-319-58961-9_12
13. Carletti, V., Foggia, P., Vento, M.: Performance comparison of five exact graph matching algorithms on biological databases. In: Petrosino, A., Maddalena, L., Pala, P. (eds.) ICIAP 2013. LNCS, vol. 8158, pp. 409–417. Springer, Heidelberg (2013). https://doi.org/10.1007/978-3-642-41190-8_44
14. Carletti, V., Foggia, P., Vento, M.: VF2 plus: an improved version of VF2 for biological graphs. In: Liu, C.-L., Luo, B., Kropatsch, W.G., Cheng, J. (eds.) GbRPR 2015. LNCS, vol. 9069, pp. 168–177. Springer, Cham (2015). https://doi.org/10.1007/978-3-319-18224-7_17
15. Coffman, T., Greenblatt, S., Marcus, S.: Graph-based technologies for intelligence analysis. Commun. ACM **47**, 45–47 (2004)
16. Conte, D., Foggia, P., Sansone, C., Vento, M.: Thirty years of graph matching in pattern recognition. Int. J. Pattern Recogn. Artif. Intell. **18**, 265–298 (2004)

17. Cordella, L., Foggia, P., Sansone, C., Vento, M.: A (sub)graph isomorphism algorithm for matching large graphs. IEEE Trans. Pattern Anal. Mach. Intell. **26**, 1367–1372 (2004)
18. Foggia, P., Percannella, G., Vento, M.: Graph matching and learning in pattern recognition on the last ten years. J. Pattern Recogn. **28**, 1450001 (2014)
19. Foster, I.: Designing and Building Parallel Programs: Concepts and Tools for Parallel Software Engineering. Addison-Wesley Longman Publishing Co., Inc., Boston (1995)
20. Jenkins, J., Arkatkar, I., Owens, J.D., Choudhary, A., Samatova, N.F.: Lessons learned from exploring the backtracking paradigm on the GPU. In: Jeannot, E., Namyst, R., Roman, J. (eds.) Euro-Par 2011. LNCS, vol. 6853, pp. 425–437. Springer, Heidelberg (2011). https://doi.org/10.1007/978-3-642-23397-5_42
21. Lacroix, V., Fernandez, C., Sagot, M.: Motif search in graphs: application to metabolic networks. Trans. Comput. Biol. Bioinf. **3**, 360–368 (2006)
22. McCreesh, C., Prosser, P.: A parallel, backjumping subgraph isomorphism algorithm using supplemental graphs. In: Pesant, G. (ed.) CP 2015. LNCS, vol. 9255, pp. 295–312. Springer, Cham (2015). https://doi.org/10.1007/978-3-319-23219-5_21
23. Paten, B., Novak, A.M., Eizenga, J.M., Garrison, E.: Genome graphs and the evolution of genome inference. Genome Res. **27**, 665–676 (2017)
24. Rodenas, D., Serratosa, F., Solé-Ribalta, A.: Parallel graduated assignment algorithm for multiple graph matching based on a common labelling. In: Jiang, X., Ferrer, M., Torsello, A. (eds.) GbRPR 2011. LNCS, vol. 6658, pp. 132–141. Springer, Heidelberg (2011). https://doi.org/10.1007/978-3-642-20844-7_14
25. Solnon, C.: Alldifferent-based filtering for subgraph isomorphism. Artif. Intell. **174**, 850–864 (2010)
26. Vento, M.: A long trip in the charming world of graphs for pattern recognition. Pattern Recogn. **48**, 291–301 (2014)
27. Wasserman, S., Faust, K.: Social Network Analysis: Methods and Applications. Cambridge University Press, Cambridge (1994)
28. Xie, X., Li, Z., Zhang, H.: Efficient subgraph matching in large graph with partitioning scheme. In: 13th Web Information Systems and Applications Conference (2016)
29. Xu, Q., Jeon, H., Annavaram, M.: Graph processing on GPUs: where are the bottlenecks. In: 2014 IEEE International Symposium on Workload Characterization (2014)

Local Binary Pattern Based Graph Construction for Hyperspectral Image Segmentation

Kaouther Tabia[✉], Xavier Desquesnes, Yves Lucas, and Sylvie Treuillet

Laboratoire PRISME, Université d'Orléans,
12 rue de Blois 6744, 45067 Orléans Cedex 2, France
kaouther.tabia@univ-orleans.fr

Abstract. Building highly discriminative graph has an important impact on the quality of graph-based hyperspectral image segmentation. For this purpose, we propose to weight graph edges using Local Binary Pattern (LBP) descriptor that takes into account the texture information of the hyperspectral images. Nodes in the graph embed spectral LBP features computed from the different hyperspectral bands, while edges encode the spatial relationship between these features.

The multiphase level set method is then applied on the constructed graph to segment the image. We validate the proposed method, using Overlapping Score evaluation metric, on several popular hyperspectral images. The results show that our method is very efficient compared to other state-of-the-art one.

Keywords: Hyperspectral image · Graph · LBP · Segmentation · Multiphase level set

1 Introduction

Hyperspectral imaging has recently gained in popularity as a promising optical image acquisition modality, after having been limited to costly remote sensing devices. Hyperspectral sensors can provide a fine sampling of the visible and near infrared spectrum. Hyperspectral data are represented in a three-dimensional image: two spatial dimension and one spectral dimension. Each image pixel corresponds to a high dimensional vector of hundreds of spectral bands captured at this pixel called spectrum. This increased spectral information may, in turn, contribute to a sharper image analysis including hyperspectral image (HSI) segmentation. The aim of HSI segmentation is to partition the image into a set of regions sharing the same properties and characteristics, which may serve as a prior step to many applications such as classification.

Over the past decade, a considerable amount of research has focused on HSI segmentation and classification [2, 4, 7].

Among proposed methods, one can notice the rise of graph-based algorithms, which have become well-established tools for HSI segmentation problems.

© Springer Nature Switzerland AG 2019
D. Conte et al. (Eds.): GbRPR 2019, LNCS 11510, pp. 152–160, 2019.
https://doi.org/10.1007/978-3-030-20081-7_15

Graph-based algorithms on the other hand rely upon the construction of a discriminant graph representation.

Moreover, the graph construction is not constrained to usual regular-grid topologies, where each pixel is connected to its adjacent neighbors, but can be adapted to each particular application. One can mention large or complete neighborhood for textured images or Region Adjacency Graphs (RAG) where pixels are replaced by regions or superpixels, and so on.

In a recent study [10] the spatial-spectral structure of weighted graphs has shown its potential for HSI segmentation with the multiphase level set method. In the aforementioned method [10], the weighting of the graph edges was limited to calculating the spectral angle mapper between the image pixels which represent the vertices in the graph.

In this paper, we propose a new metric for edges weighting based on the generalization of the LBP feature on graphs.Our idea is inspired from the Local Binary Pattern (LBP) which was originally proposed by [8] for texture analysis and has later been used in many fields including visual inspection [5], face recognition [6] and motion analysis [1].

LBP is a non-parametric descriptor whose aim is to efficiently summarize the local structures of images. Its advantage over other approaches are its simplicity and effectiveness. This motivated us to propose a novel graph construction based on LBP features. Our aim is to capture the dominant features of the vertices with their neighbors and to encode the local structure around each vertex, before to obtain a small set of the most discriminative LBP-based features for better performance.

2   Preliminaries

In this section we give some basic definitions of important terminologies which are used throughout this paper.

2.1   Graph

A graph $G = (V, E)$ consists of a finite nonempty set of vertices $V = (v_1, ..., v_n)$, and a finite set of edges $E = \{(u, v) \in V \times V | u \sim v\}$ where $\sim$ means that u and v are adjacent vertices, and a weight function, denoted $\omega: V \times V \to [0, 1]$, which represents the weight of each edge in G corresponding to the amount of interaction between two vertices.

In the rest of this paper we denote $\omega(u, v)$ by ω_{uv}. By convention, this function verifies the following properties:

$$\omega_{uv} \begin{cases} \in [0, 1] \; \forall (u, v) \in E \\ = 0 \; \forall (u, v) \notin E \\ = \omega_{vu} \; \text{symmetry} \end{cases} \tag{1}$$

We now review some operators on weighted graphs [3]. For a given discrete function f which assigns a real value $f(u)$ to each vertex $u \in V$, the weighted discrete partial derivative operator of f applied on an edge $(u, v) \in E$ is:

$$\partial_v f(u) = \sqrt{\omega_{uv}} \left(f(v) - f(u) \right) \tag{2}$$

Based on this definition, two weighted directional difference operators are defined. The external and internal difference operators are respectively:

$$\partial_v^+ f(u) = \sqrt{\omega_{uv}} \left(f(v) - f(u) \right)^+ \tag{3}$$

and

$$\partial_v^- f(u) = -\sqrt{\omega_{uv}} \left(f(v) - f(u) \right)^- \tag{4}$$

with $(x)^+ = \max(0, x)$ and $(x)^- = \min(0, x)$. The weighted gradient of f at vertex u, denoted ∇_w is the vector of all edge directional derivatives:

$$(\nabla_w f)(u) = \left(\partial_v f(u) \right)_{(u,v) \in E}^T \tag{5}$$

The external and the internal weighted gradient operators of f, denoted respectively ∇_w^+ and ∇_w^- are:

$$(\nabla_w^\pm f)(u) = \left(\partial_v^\pm f(u) \right)_{(u,v) \in E}^T \tag{6}$$

2.2   Local Binary Patterns

The LBP feature has originally been introduced by Ojala et al. [8] for 2D texture analysis. LBP is a non-parametric approach, which accurately summarizes local image structure by comparing central pixels with their neighbors, encoding their relations using a binary code.

Formally, the original LBP operator [8] computes binary codes of image pixels by thresholding the 3×3 neighborhood of each pixel with the center value in a clockwise rotation starting from the top-left one. Pixel neighbors whose intensities are greater or equal to the central pixel's are marked as 1, otherwise as 0. The resulting sequence of 0 and 1 is considered as a 8-digit binary number. Converting this binary sequence into a decimal number we obtain the LBP code of the central pixel.

Figure 1 summarizes the different steps of the LBP calculation. In the context of texture classification using LBP method, the occurrences of the codes (decimal values of LBP codes) in an image are collected in a histogram. The classification is then performed by a simple calculation of distance between histograms.

One limitation of the basic LBP operator is that its small 3×3 neighborhood cannot capture dominant features with large scale structures. To deal with the texture at different scales, the operator was later generalized to use neighborhoods of different sizes [9]. A local neighborhood is defined as a set of sampling points evenly spaced on a circle which is centered at the pixel to be labeled, and the sampling points that do not fall within the pixels are interpolated using bilinear interpolation, thus allowing for any radius and any number of sampling points in the neighborhood.

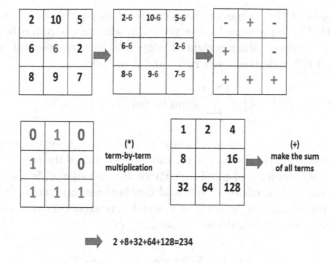

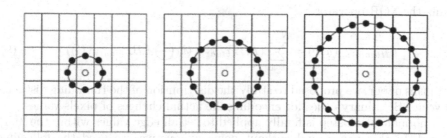

2 +8+32+64+128=234

Fig. 1. Overview of LBP computing process

Depending on the scale of the neighborhood used, some regions of interest such as corners or edges can be detected by this descriptor.

Figure 2 shows some examples of the extended LBP operator, where the notation (N, R) denotes a neighborhood of N sampling points on a circle of radius of R.

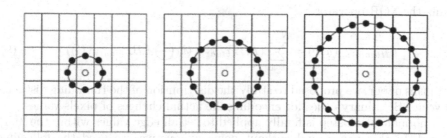

Fig. 2. Examples of the extended LBP operator: the circular (8, 1), (16, 2), and (24, 3) neighborhoods

3   LBP-based Graph Construction

In this section we illustrate the graph construction method using LBP features for a hyperspectral image. The first step consists in representing a HSI (X, Y, Λ), where X, Y and Λ represent respectively the number of lines, columns and spectral bands, as a set of overlapping patches $\{P_i, i = 1...m\}$, each patch P_i being extracted around a representative pixel $p_i = (x_i, y_i, \lambda_i)^t$.

The m pixels of the image are the centers of m patches of radius R. Each pixel in the HSI is represented by a vertex in the graph connected to the 4 nearest spatial vertices with a weighted edge. For each representative pixel p_i, we compute a LBP code over each HSI band λ as follows:

$$LBP_{\lambda_i}(p_i) = \sum_{n=0}^{N-1} sign(S_{\lambda_i}(n) - S_{\lambda_i}(c)) \times 2^n \qquad (7)$$

where $S_{\lambda_i}(c)$ represent the spectral value on the band λ_i of the central pixel, $S_{\lambda_i}(n)$ represent the spectral value on the same band of its neighbors and N represent the total number of pixel in a Patch P_i, and sign(x) is a sign function defined by $sign(x) = 1$ where $x \geq 0$ and 0 otherwise. So, the $LBP(p_i)$ (8) is a vector feature which can be seen as a serial concatenation of standard LBP codes computed over each HSI band separately (7).

$$LBP(p_i) = \sum_{k=0}^{\Lambda} LBP_{\lambda_k}(p_i) \times 2^{k \times N} \qquad (8)$$

Once these LBP vector codes are calculated for each representative pixel (i.e vertex in the constructed graph), the Hamming distance is used to compute the weights of the graph edges between each vertices pair (u,v). This is performed by applying the XOR operator between the binary chains of the LBP vector codes of corresponding vertices, which gives as results the number of positions at which the binary chains are different. Figure 3 and Eq. (9) explains the Hamming distance calculation between the two LBP binary chains of u and v, where $\oplus$ means the XOR operator.

$$d_{Hamming}(u,v) = \sum_{i=0}^{N \times \Lambda} (LBP_{binary_u}(i) \oplus LBP_{binary_v}(i)) \qquad (9)$$

In this paper, we proposed to study the distribution of the Hamming distance between LBP binary codes for encoding the actual changes of pixels values.

A normalization step is finally applied to set all edge values within the $[0, 1]$ interval. In this paper, and without being exhaustive, we used the following normalization (10), where σ represents the standard deviation calculated over all Hamming distances:

$$\omega(u,v) = exp(-\frac{d_{Hamming}(u,v)^2}{\sigma^2}) \qquad (10)$$

4   HSI Segmentation

After HSI graph construction and calculating W affinity matrix which represents the weight of all the edges in the graph, we apply the multiphase level set graph based method [10] to partition the graph into n regions and deduce HSI segmentation.

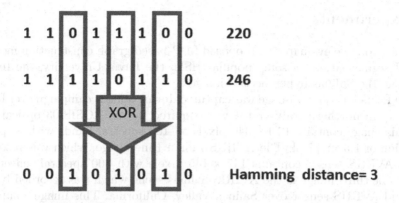

Fig. 3. Hamming distance calculation between LBP codes

We first initialize the n level set functions ϕ_i by n initial contours on the image, then we compute the averages in each region composed by the superposition of these contours and we calculate the speed propagation function F_n to solve the curve evolution equation as follows:

$$\frac{\partial \phi_n(u,t)}{\partial t} = \begin{cases} F_n(u,t) \, ||(\nabla_\omega^! \phi_n)(u,t)||, & \text{if } F_n(u,t) > 0 \\ F_n(u,t) \, ||(\nabla_\omega^- \phi_n)(u,t)||, & \text{if } F_n(u,t) < 0 \\ 0, & \text{otherwise} \end{cases} \tag{11}$$

where $(\nabla_\omega^! \phi_n)(u,t)$ and $(\nabla_\omega^- \phi_n)(u,t)$ are the external and internal operators of weighted gradients. Based on Eqs. (3), (4) and (6), the external and internal operators of weighted gradients of ϕ_n can be defined by:

$$(\nabla_\omega^\pm \phi_n)(u) = \pm\sqrt{\omega_{uv}}((\phi_n(v) - \phi_n(u))^\pm)_{(u,v)\in E}^T \tag{12}$$

Finally we iterate the last three steps. We show in Fig. 4 an example of initial contours using the four-phase (with two level set functions) models.

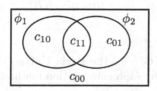

Fig. 4. Two contours split the image into 4 regions: $c_{11} = \{\phi_1 > 0, \phi_2 > 0\}, c_{10} = \{\phi_1 > 0, \phi_2 < 0\}, c_{01} = \{\phi_1 < 0, \phi_2 > 0\}, c_{00} = \{\phi_1 < 0, \phi_2 < 0\}$

5   Experiments

In this section, we evaluate the proposed LBP-based graph construction method for HSI segmentation on some popular HSIs: the Pavia University, the Indian Pines and the Salinas hyperspectral images.

The Pavia University image was captured during a flight campaign over Pavia university in northern Italy and it was acquired by the ROSIS-03 optical sensor. This image contains 610 × 340 pixels on 103 spectral bands with a spatial resolution of 1.3 m/pixel. The AVIRIS Indian Pine image, which was recorded by the AVIRIS sensor contains 145 × 145 pixels with 200 spectral reflectance bands. The third image is the AVIRIS Salinas scene which was recorded by the 224-band AVIRIS sensor over Salinas Valley, California. This image comprises 512 × 217 pixels with a high spatial resolution of 3.7 m/pixel.

We evaluated the proposed LBP-based graph construction method with a patch size 7 × 7 corresponding to a circular (24, 3) neighborhoods (Fig. 2). We compared our construction graph approach with the 4-neighborhood graph construction using the multiphase level set segmentation proposed in [10].

Figure 5 represent a zoom on some different regions taken from our HSIs and the segmentation results in this regions presented in line-wise, from top to bottom, line 1 and 4 shows the RGB original images, line 2 and 5 shows the segmentation results using the four-phase model with the proposed LBP-based graph construction, when line 3 and 6 show the segmentation results with the aforementioned method [10].

From this visual comparison of our results with the other method, we can see that our proposed method can obtain more meaningful region with accurate boundary. It can be observed that our method can well segment the texture images (the meadows, the brick) and it has high discriminative power to detect objects from different backgrounds.

To evaluate the results obtained for HSIs segmentation we compute the Overlapping Score (OS) to compare the segmented image S and the ground truth G. The OS metric is defined by:

$$OS = \frac{|S \cap G|}{min(|S|, |G|)} \tag{13}$$

Table 1. Overlapping Score (OS) for the Pavia University, Indian Pines and Salinas images segmentation with both graph construction methods (four-phase segmentation)

HSI	Method	
	4-neighborhood graph construction	LBP-based graph construction
Pavia University	0.8387	**0.9102**
Indian Pines	0.8563	**0.9275**
Salinas	0.8392	**0.9134**

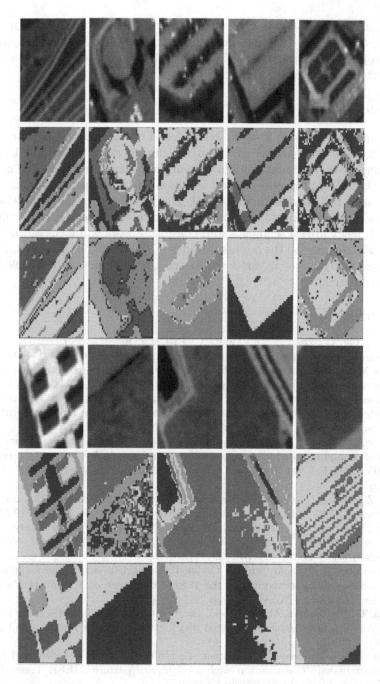

Fig. 5. Visual comparison of our results with the other method (Lines (1 and 4) RGB regions of interest extracted from Pavia University, Indian Pines and Salinas images, lines (2 and 5) the four-phase level set segmentation results on LBP-based graph construction, lines (3 and 6) the four-phase level set segmentation on 4-neighborhood graph construction)

Evaluation of segmentation results is given in Table 1, with the best results highlighted in bold for each measurement. It is obvious to see that the proposed LBP-base graph construction method ranks the first place compared with the other method.

6   Conclusion

In this paper, we propose a new method based on the LBP descriptor to construct the graph for hyperspectral image segmentation. The proposed method is invariant to uneven light conditions and noise benefiting from the usage of LBP patches. The proposed method is evaluated by extensive experiments on three popular HSI, and is quantitatively compared with some other standard algorithms. The experimental results have shown the potential of our method and its efficiency in HSI segmentation. Our future work will focus on the effects of the LBP patch size on the segmentation performance of our proposed method, we will also analyze how it varies with respect to the number of sample points in the LBP patch.

References

1. Cai, L., Ge, C., Zhao, Y.M., Yang, X.: Fast tracking of object contour based on color and texture. Int. J. Pattern Recogn. Artif. Intell. **23**(07), 1421–1438 (2009)
2. De La Vega, S.H., Manian, V.: Object segmentation in hyperspectral images using active contours and graph cuts. Int. J. Remote Sens. **33**(4), 1246–1263 (2012)
3. Desquesnes, X., Elmoataz, A., Lézoray, O.: Eikonal equation adaptation on weighted graphs: fast geometric diffusion process for local and non-local image and data processing. J. Math. Imaging Vis. **46**(2), 238–257 (2013)
4. Dópido, I., Villa, A., Plaza, A., Gamba, P.: A quantitative and comparative assessment of unmixing-based feature extraction techniques for hyperspectral image classification. IEEE J. Sel. Top. Appl. Earth Observations Remote Sens. **5**(2), 421–435 (2012)
5. Guo, Z., Zhang, L., Zhang, D.: Rotation invariant texture classification using LBP variance (LBPV) with global matching. Pattern Recogn. **43**(3), 706–719 (2010)
6. Huang, D., Shan, C., Ardabilian, M., Wang, Y., Chen, L.: Local binary patterns and its application to facial image analysis: a survey. IEEE Trans. Syst. Man Cybern. Part C **41**(6), 765–781 (2011)
7. Massoudifar, P., Rangarajan, A., Gader, P.: Superpixel estimation for hyperspectral imagery. In: Proceedings of the IEEE Conference on Computer Vision and Pattern Recognition Workshops, pp. 287–292 (2014)
8. Ojala, T., Pietikäinen, M., Harwood, D.: A comparative study of texture measures with classification based on featured distributions. Pattern Recogn. **29**(1), 51–59 (1996)
9. Ojala, T., Pietikainen, M., Maenpaa, T.: Multiresolution gray-scale and rotation invariant texture classification with local binary patterns. IEEE Trans. Pattern Anal. Mach. Intell. **24**(7), 971–987 (2002)
10. Tabia, K., Desquesnes, X., Lucas, Y., Treuillet, S.: A multiphase level set method on graphs for hyperspectral image segmentation. In: Blanc-Talon, J., Distante, C., Philips, W., Popescu, D., Scheunders, P. (eds.) ACIVS 2016. LNCS, vol. 10016, pp. 559–569. Springer, Cham (2016). https://doi.org/10.1007/978-3-319-48680-2_49

A Parallel MCMC Algorithm
for the Balanced Graph Coloring Problem

Donatello Conte[1], Giuliano Grossi[2], Raffaella Lanzarotti[2], Jianyi Lin[3(✉)],
and Alessandro Petrini[2]

[1] Université de Tours, Computer Science Laboratory (LIFAT - EA6300),
64 Avenue Jean Portalis, 37000 Tours, France
donatello.conte@univ-tours.fr
[2] Dipartimento di Informatica, Università degli Studi di Milano,
Via Celoria 18, 20133 Milan, Italy
{grossi,lanzarotti}@di.unimi.it, alessandro.petrini@unimi.it
[3] Department of Mathematics, Khalifa University of Science and Technology,
Al Saada St., PO Box 127788, Abu Dhabi, United Arab Emirates
jianyi.lin@ku.ac.ae

Abstract. In parallel computation domain, graph coloring is widely
studied in its own and represents a reference problem for scheduling of
parallel tasks. Unfortunately, common graph coloring strategies usually
focus on minimizing the number of colors without any concern for the
sizes of each color class, thus producing highly skewed color class distributions. However, to guarantee efficiency in parallel computations, but
also in other application contexts, it is important to keep the color classes
highly balanced in their sizes. In this paper we address this challenging
issue for large scale graphs, proposing a fast parallel MCMC heuristic
for sparse graphs that randomly generates good balanced colorings provided that a sufficient number of colors are made available. We show its
effectiveness through some numerical simulations on random graphs.

Keywords: Balanced graph coloring ·
Markov Chain Monte Carlo method · Greedy colorer ·
Parallel algorithms

1 Introduction

The vertex coloring (or graph coloring) problem is one of the fundamental and
most difficult combinatorial problems. Given an undirected graph G, one is looking for an assignment of colors to the vertices of G such that no two adjacent
vertices share the same color and the number of different colors used is minimized. Vertex coloring is known to be NP-hard even for planar graphs [7]. Graph
coloring has many applications in different research fields. For example, it can be
used to register medical or biometric images [4,16] and to find good resource allocation scheme for device-to-device (D2D) communications [3,22], used in modern wireless communication systems [11]. Moreover, graph coloring is extensively

© Springer Nature Switzerland AG 2019
D. Conte et al. (Eds.): GbRPR 2019, LNCS 11510, pp. 161–171, 2019.
https://doi.org/10.1007/978-3-030-20081-7_16

used in social networks problem such as Community Identification in Dynamic Social Networks [21], summarization of social networks messages [19], and for Collective Spammer Detection [6]. A common characteristic of these tasks is that the graphs have very large size, thus requiring a speed up of the traditional greedy sequential coloring heuristics [2], obtained introducing parallelization. In Patter Recognition, also, there are many applications of graph coloring, that are intractable without efficient parallelization: e.g. stochastic pyramids construction [17], graph classification [9], and so on.

In the literature, parallel graph coloring problem has been tackled by several approaches but, at the best of our knowledge, very few address the problem of balancing in parallel manner. One category of them is based on the search of a maximal independent set of vertices on a progressively shrunk graph and the concurrent coloring of the vertices in the found independent set. Often the independent set itself is computed in parallel using some variant of the Luby's algorithm [15]. Examples of such approaches are [8,13]. Another category includes methods that color as many vertices as possible concurrently, tentatively tolerating potential conflicts, while detecting and solving conflicts afterwards (e.g. [1]). Despite these solutions are effective in producing a proper coloring, generally minimizing the number of colors, they produce highly skewed color classes, undesirable for many applications, such as parallel job scheduling, that requires balancing among the classes. At the other extreme, one could search for a coloring being *equitable*, that is a coloring that guarantees that the sizes of any two color classes differ by at most one [18]. This constraint is very expensive and somehow too stringent for practical applications. *Balanced* coloring relaxes the equitable constraint requiring that any two color class sizes differ by an integer l greater than 1. Few approaches have been proposed to tackle Balanced graph coloring (e.g. [14,20]). However, the limit of these methods is still that they are intrinsically sequential thus not scalable, becoming unfeasible on large graph.

A promising direction of research on graph coloring concerns the Markov Chain Monte Carlo (MCMC) methods that allow sampling from non analytic complex distributions. The idea is to define an ergodic Markov chain whose steady state distribution is defined over the set of colorings we wish to sample from. Within the framework of graph coloring using Markov chains several contributions have been proposed. In [12] a simple sequential solution based on the Glauber dynamics has been adopted. The Glauber dynamics produces a Markov chain on a proper coloring where at each step a random vertex v is recolored, choosing a color uniformly at random from the permissible ones.

In this article we present an algorithm based on MCMC method producing balanced graph coloring in a parallel way. The main contribution is the introduction of a proposal distribution, independently for each vertex, that promotes overall balancing objectives. The key computational property is that the generation of colorings with such distributions can be carried out on parallel computational models. The technique is tested on large random graphs showing experimentally its effectiveness.

The remainder of the paper is organized as follows: Sect. 2 describes the proposed algorithm and in Sect. 3 we prove quantitatively, by some experimental results, the effectiveness of our method.

2 Parallel MCMC Sampling

2.1 Notations

Let $G = \langle V, E \rangle$ be a simple undirected graph of $n = |V|$ vertices and $[k] = \{1, \ldots, k\}$ be a set of colors used to label the vertices. A k-coloring is an assignment $c : V \to [k]$ such that $c = (c(1), \ldots, c(n)) \in [k]^n$ is called *proper* if adjacent vertices receive different colors, otherwise it is termed *improper*. It is well-known that, if $\Delta(G)$ is the maximum degree of G, $k = \Delta(G) + 1$ colors are sufficient to properly color the graph by a sequential greedy algorithm. For a given coloring c, let $\mathcal{N}(v)$ denote the neighborhood of node v in G, and $c_{\mathcal{N}(v)} \subseteq [k]$ be the set of colors occupied by vertices $\mathcal{N}(v)$ and $\bar{c}_{\mathcal{N}(v)}$ its complement. The neighborhood of v induces the partition $\{c_{\mathcal{N}(v)}, \bar{c}_{\mathcal{N}(v)}\}$ of $[k]$ for which $h_v(c) = |c_{\mathcal{N}(v)}|$ and $\bar{h}_v(c) = |\bar{c}_{\mathcal{N}(v)}|$ denote their cardinality. We will also consider the absolute *frequency* of the color j in c: $f_j(c) = |\{u \in V : c_u = j\}|$.

Hereafter we will use lowercase letters, e.g. c, c', c^*, for given colorings and uppercase for random colorings, e.g. C, C', C^*. For example the probability of $C' = c'$ given $C = c$ will be denoted $\mathbb{P}(C' = c' \mid C = c)$ or $\mathbb{P}(c' \mid c)$ for short.

2.2 Markov Chain Monte Carlo for Sampling Colorings

Monte Carlo estimation methods [23] are a broad class of statistical sampling techniques based on the idea of estimating an unknown quantity with averaging over a large set of samples. Due to the strong law of large numbers, the estimate is guaranteed to almost surely converge to the unknown quantity. When the iterative sampling scheme is based on the distribution of a Markov Chain it is termed Markov Chain Monte Carlo (MCMC). Due to the lack of space here, we refer to [23] for a comprehensive introduction to the topic.

Our objective is to define a 1st-order ergodic Markov chain $(C_i)_{i=1}^{\infty}$ consisting in a sequence of k-colorings of a simple undirected graph $G = \langle V, E \rangle$ with $n = |V|$ nodes, whose stationary distribution π strongly depends on the set of conflicts (edges with endpoints sharing the same color) involved in each coloring $c \in [k]^n$. A natural target stationary distribution for the Markov chain is the *Gibbs distribution*, expressed in the form

$$\pi(c) = \frac{e^{-\beta \#(c)}}{Z(\beta)}, \quad \text{with } Z(\beta) = \sum_{c' \in [k]^n} e^{-\beta \#(c')}, \tag{1}$$

where $\#(c) : [k]^n \to \mathbb{N}$ counts the number of conflicts of the coloring c.

The choice of Gibbs distribution (1) turns out to be useful since, when β is sufficiently large, it is close uniformly and with exponential rate to the uniform distribution over the *proper* colorings, which is our desired working set.

It is known from MCMC theory that the latter distribution is asymptotically approached in the sampling process when the chain is suitably constructed using the established Metropolis-Hastings algorithm [10]. This technique prescribes the specification of a *proposal* probability encapsulating the *acceptance ratio* and a *transition* probability for the chain.

As for the transition probabilities of the Markov chain, given a coloring $C = c$ we sample the successive coloring C^* in two phases acting according to a typical "rejection sampling" scheme [23]: first a candidate coloring C' is generated according to a suitable proposal probability $r(c, c') := \mathbb{P}(C' = c' \mid C = c)$, then the proposal C' is accepted effectively as successive coloring C^* according to the acceptance ratio $\alpha(c, c')$:

$$\mathbb{P}(C^* = c' \mid C = c) = \alpha(c, c') := \min \left\{ \frac{\pi(c')r(c', c)}{\pi(c)r(c, c')}, 1 \right\}, \qquad \text{where } c' \neq c$$

while the old coloring $C = c$ is retained with remaining probability $1 - \alpha(c, c')$.

The proposal coloring C' is sampled with probability $r(c, c')$ as follows. Each node $v \in V$ is drawn independently and with identical distribution $\mathbb{P}(c'_v \mid c)$ of colors so that the overall proposal probability is

$$r(c, c') = \prod_{v \in V} \mathbb{P}(c'_v \mid c). \tag{2}$$

Notice that, in the construction of the acceptance ratio also the *backward* probability $r(c', c)$ is required, hence $r(c, c')$ is called *forward* probability.

The choice of the node proposal probability $\mathbb{P}(c'_v \mid c)$ is a key step and is hence detailed distinctly in the following subsection. It is also important to observe from the computational viewpoint that the independent drawing of all c'_v, $v \in V$, allows for the generation of the new coloring in a parallel manner.

2.3   Generation of Proposal Color

Here we specify the proposal probability in algorithmic vein so that the stochastic evaluations follow consequently from the analysis of the color generation. First, the behavior of the algorithm splits into two cases based on the old coloring c. When there is some conflict locally for v, namely $c_v \in c_{\mathcal{N}(v)}$, the new proposed color C'_v for v shall be redrawn with the aim of reducing the possible conflicts. We draw it from the *available* colors $\bar{c}_{\mathcal{N}(v)}$ following a nearly uniform distribution of $C'_v = j$ given c:

$$\eta_v(j, c) = \begin{cases} \frac{1 - \varepsilon h_v(c)}{k - h_v(c)} & \text{if } j \in \bar{c}_{\mathcal{N}(c)} \\ \varepsilon & \text{if } j \in c_{\mathcal{N}(c)}. \end{cases}$$

The rationale behind such definition is that we want with high probability to generate a color equally likely among those available, in order to aim at the balancing objective of the method. Nevertheless, we keep a negligible chance $\varepsilon > 0$ to pick a color that is not available, in order to widen the search space. In the clearly rare case of no available colors, the algorithm maintains the old color c_v with high probability, $1 - \varepsilon(n-1)$, and selects another color with small probability ε.

As for the case of no conflict for v, $c_v \in \bar{c}_{\mathcal{N}(c)}$, it is desirable to keep the old color c_v nearly surely to facilitate the convergence of the algorithm, or otherwise pick another color with a small chance ε. Hence, for the node v the proposal color $C'_v = j$ given c in the case of no conflict is distributed as

$$\zeta_v(j, c) = \begin{cases} 1 - \varepsilon(n-1) & \text{if } j = c_v \\ \varepsilon & \text{if } j \neq c_v. \end{cases}$$

So far, we have presented the elements for determining the forward probability $r(c, c')$ in (2). Indeed, from the discussion above one can derive the following

Proposition 1. *The proposal probability of each node v is*

$$\mathbb{P}(c'_v \mid c) = \begin{cases} \eta_v(c'_v, c) & \text{if } h_v(c) < k, c_v \in c_{\mathcal{N}(c)} \\ \zeta_v(c'_v, v) & \text{otherwise.} \end{cases}$$

On the other hand, the backward probabilities

$$r(c', c) = \mathbb{P}(C' = c \mid C = c') = \prod_{v \in V} \mathbb{P}(C'_v = c_v \mid C = c')$$

can be obtained by symmetrical reasoning, i.e. exchanging the role of c and c' in the calculations outlined above. This allows to compute then the acceptance ratio $\alpha(c, c')$.

2.4   Algorithm

We can now describe more formally the procedural steps corresponding to our method providing Algorithm 1. As for the convergence properties of the proposed

---

Algorithm 1. Parallel MCMC Balanced Graph Coloring

Input: Graph $G = \langle V, E \rangle$ with $n = |V|$; Number k of colors; Gibbs parameter $\beta \ll 1$
Output: Random proper coloring $C \in [k]^n$
1: $C :=$ some initial arbitrary coloring $\in [k]^n$
2: **while** $\#(C) > 0$ **do**
3:    **for each** $v \in V$ **in parallel do**
4:        Calculate $C_{\mathcal{N}(C)}$ and $h_v(C) := |C_{\mathcal{N}(C)}|$
5:        Compute $\mathbb{P}(c'_v \mid C)$ according to the rule in Prop. 1
6:        Generate proposal color C'_v with distribution $\mathbb{P}(c'_v \mid C)$
7:    Proposed coloring $C' := (C'_1, C'_2, ..., C_n)$
8:    Compute forward and backward probabilities $r(C, C')$, $r(C', C)$
9:    $\alpha(C, C') := \min\{\frac{r(C', C)}{r(C, C')} e^{-\beta(\#(C') - \#(C))}, 1\}$
10:    Accept proposed coloring, $C := C'$, with probability $\alpha(C, C')$
11: **return** C

algorithm, since we cannot guarantee that the number of conflicts $\#(C)$ strictly decreases at each iteration, due to the randomness of the MCMC methodology, we are able to give a characterization of the convergence in stochastic terms as in the following proposition, which is stated without proof for lack of space.

Proposition 2. *Let C^t and C^{t+1} be the random coloring in iteration t and $t+1$, respectively, of Algorithm 1 on G. Provided a number of colors $k \geq \Delta(G) + 1$, it holds the expectation inequality:*

$$\mathbb{E}[\#(C^{t+1})] < \mathbb{E}[\#(C^t)].$$

It follows that, the number of colors $\#(C)$ in the algorithm converges in probability to 0, i.e. $\lim_{t \to \infty} \mathbb{P}(\#(C^t) < \theta) = 1 \; \forall \theta > 0$.

3   Numerical Simulations

In this section we report some numerical simulation results of the parallel MCMC method exploiting the Erdős-Rènyi graph (ER) model [5]. This model is widely used to generate random graphs and has some valuable properties to leverage in order to asses the behaviour of the proposed coloring strategy both for fixed graph sizes and asymptotically.

In the ER model $G(n, p)$, a n-vertex graph is constructed by connecting vertices randomly and including each edge with probability p independently from every other edge. Equivalently, the probability that a vertex v has degree k is Binomial, i.e. $\mathbb{P}(\deg(v) = k) = \binom{n-1}{k} p^k (1-p)^{n-1-k}$, with expected value $E[\deg(v)] = (n-1)p$. As n goes to infinity, the probability that a graph in $G(n, 2\ln(n)/n)$ is connected, tends to one. Another relevant property of ER graphs is the edge density which is a random variable with expectation exactly equal to the background probability p.

The role that ER model plays in this picture is important because it ensures the possibility to provide graphs with certain properties, giving us at same time an effective and sound algorithmic procedure to compute them in practice.

To compare our MCMC strategy with other techniques designed for the same purpose, we choose a fast parallel greedy algorithm inspired by Luby's work [15]. Luby has described a greedy parallel strategy to find a maximal independent set (MIS) of vertices (i.e. a subset of vertices such that no two vertices are neighbors) in undirected graphs. Consequently, given that any MIS can be colored in parallel, a greedy graph coloring strategy could be defined by repeatedly finding the largest MIS on subgraphs gradually resulting from pruning previous recovered MIS.

Clearly, the Luby inspired colorer is not meant for balanced graph coloring problem. However, we use it for two reasons: on one hand just for sake of comparison with a simple scheme graph colorer, on the other hand to empirically show that the two algorithms have comparable computational times on sparse graphs.

3.1   CUDA Parallel Implementations

We developed a fast parallel implementation of both the MCMC and the Luby-greedy coloring algorithms, called respectively MCMC-GPU and Luby-GPU, using the NVIDIA CUDA programming paradigm. NVIDIA GPU processors feature up to 5000 processing cores, hence a very large number of processing threads can be scheduled and executed concurrently in a shared memory model. Thanks to this, parallelization in both MCMC-GPU and Luby-GPU occurs at vertex level, i.e. a thread is assigned to each vertex of the graph which is therefore processed concurrently to every other vertex.

MCMC-GPU implementation closely follows Algorithm 1: during the iterations, each vertex v is assigned to a processing thread which evaluates both $\mathbb{P}(c_v' \mid c)$ and $\mathbb{P}(c_v \mid c')$ (the forward and backward probabilities) and draws the new color c_v accordingly. Thread synchronization occurs only at the end of each iteration, where the total number of conflicts and the rejection factor $\alpha(c, c')$ of the new coloring have to be evaluated. In MCMC-GPU the algorithm is executed until a proper k-coloring is found, or the maximum number of allowed iterations is reached.

Also in Luby-GPU parallelization occurs at vertex level: each vertex is assigned to a thread that evaluates its status (free or not-free) and randomly adds itself to the current MIS if and only if it does not generate any conflict. Synchronization is more invasive, since it has to be performed in every stage of the MIS assembly.

3.2   Performances on ER Graphs

Here we report some preliminary experimental results on ER graphs comparing the MCMC-GPU and Luby-GPU algorithms running on NVIDIA GPU devices.

We aim at measuring the quality of balancing in the color class sizes produced in particular by MCMC-GPU, which works by improving the balancing of an existing (improper) coloring moving vertices from one color class to another on the basis of random choices, as highlighted by the proposed distribution over colors given in Sect. 2.3. Let us first introduce a measure of deviation of a coloring $c \in [k]^n$ from a perfectly balanced coloring where each color class j has size $n_j = n/k$, for each $j \in [k]$. This can be quantified by defining $\gamma_{n,k}(j) = |f_j - n/k|$ which represents the target to be minimized in respect of which our heuristic will perform local random arrangements. An overall measure of balancing quality closely related to the standard deviation of the color class sizes can be then defined as

$$\Gamma_{n,k}(c) = \left(\frac{1}{k} \sum_{j=1}^{k} \gamma_{n,k}^2(j) \right)^{1/2} \tag{3}$$

called *unbalancing index* hereafter. Clearly, a coloring c is perfectly balanced if $\Gamma_{n,k}(c) = 0$.

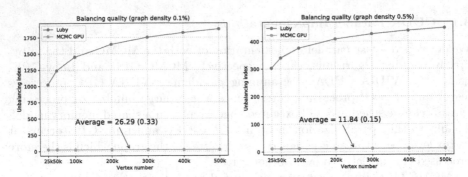

Fig. 1. Average unbalancing index achieved by MCMC-GPU and Luby-GPU on ER graphs of various size and densities 0.1% and 0.5% respectively.

A demonstration on how MCMC-GPU and Luby-GPU perform in terms of color balancing is given in Fig. 1, where the curves represent the unbalancing index (3). More precisely, the graphics relate to average amounts achieved on ER graphs of various size and densities 0.1% and 0.5% respectively. To capture a wide scale of ER graphs, once the density p has been fixed, we varied the graph size n up to 500K vertices, averaging over 10 trials for each pair (n,p) to reach satisfactory confidence level.

Concerning the number of colors k used by MCMC-GPU, assuming a regular structure of the generated graphs we fix $k = \lceil np \rceil$ corresponding to the expected vertex degree, which anyway assures the existence of proper coloring with high probability. Note that, under this setting a coloring c for a graph in $G(n,p)$ has balancing index

$$\Gamma_{n,\lceil np \rceil}(c) \approx \left(\frac{1}{np} \sum_{j=1}^{k} \gamma_{n,k}^2(j) \right)^{1/2}.$$

As can be noticed in the plots, MCMC-GPU not only outperforms Luby-GPU, but also provides invariant unbalancing index with respect to the graph sizes, being $\Gamma_{n,\lceil np \rceil}(c) \approx$ constant (with very low standard deviation), for all n in the range $25K \div 500K$. In spite of the limited number of experiments, as a general trend we have that the sum of within-class deviations $\gamma_{n,k}^2(j)$ roughly grows linearly with the number of vertices, i.e. $\sum_{j=1}^{k} \gamma_{n,k}^2(j) \approx p c_p n$, where c_p is a constant depending on the density p. For instance, in Fig. 1, the constants $c_p = 26.29$ and $c_p = 11.84$ are shown for $p = 0.1\%$ and $p = 0.5\%$ respectively.

With regard to the computational times of the conducted experiments, their averages are reported in Fig. 2. Whereas we can notice very high speedups (up to 20) between sequential and parallel MCMC implementations, the times spent by MCMC-GPU and Luby-GPU remain comparable (they turn in favor of Luby-GPU only for 500K).

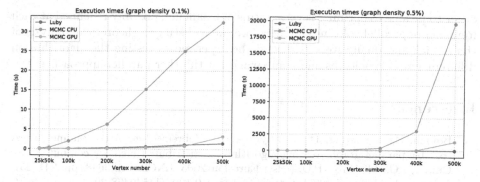

Fig. 2. Average execution times of Luby-GPU, MCMC-CPU and MCMC-GPU on ER graphs of various size and densities 0.1% and 0.5% respectively.

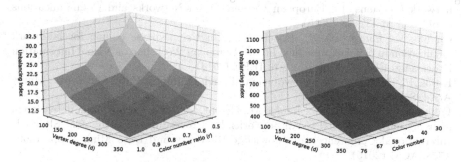

Fig. 3. Average unbalancing index achieved by MCMC-GPU (left) and Luby-GPU (right) on ER graphs varying both vertex degrees $d = \lceil np \rceil$ and color number k.

A second experiment is aimed at studying the balancing quality achieved when varying both the graph density and the number of colors made available to MCMC-GPU. In particular, here we set the graph density in terms of vertex degree $d = \lceil np \rceil$, with d falling in the range $100 \div 350$, while the number $k = r \lceil np \rceil$ of colors is scaled down by a factor $r \in (0, 1)$ ranging from 0.5 and 1. Average values of the unbalancing index over colorings carried out by the two algorithms are plotted in Fig. 3 (note that the ratio between the scales of the two graphs is about 33).

4   Conclusions

In this paper we have proposed a new parallel algorithm for graph coloring problem based on Markov Chain Monte Carlo techniques. The main goal of this new method is to produce balanced solutions, which is a direction not much explored yet in the literature. Experiments show the effectiveness of the approach on random graphs. Given the encouraging results shown in this article, further investigations are deserved. In particular we intend to generalize the model in order to be able to optimize several forms of balancing, possibly redefined at each iteration of parallel coloring; furthermore, we will study a theoretical analysis of the model, and finally we will extend the experiments testing the algorithm on different graph typologies and comparing it with other parallel approaches.

References

1. Boman, E.G., Bozdağ, D., Catalyurek, U., Gebremedhin, A.H., Manne, F.: A scalable parallel graph coloring algorithm for distributed memory computers. In: Cunha, J.C., Medeiros, P.D. (eds.) Euro-Par 2005. LNCS, vol. 3648, pp. 241–251. Springer, Heidelberg (2005). https://doi.org/10.1007/11549468_29
2. Coleman, T.F., Moré, J.J.: Estimation of sparse Jacobian matrices and graph coloring problems. SIAM J. Numer. Anal. **20**(1), 187–209 (1983)
3. Comi, P., et al.: Hardware-accelerated high-resolution video coding in virtual network functions. In: European Conference on Networks and Communications (EuCNC), pp. 32–36 (2016)
4. Cuculo, V., Lanzarotti, R., Boccignone, G.: Using sparse coding for landmark localization in facial expressions. In: 5th European Workshop on Visual Information Processing, EUVIP 2014, pp. 1–6. IEEE (2014)
5. Erdős, P., Rènyi, A.: On the evolution of random graphs. Publ. Math. Inst. Hung. Acad. Sci. **5**, 17–61 (1960)
6. Fakhraei, S., Foulds, J., Shashanka, M., Getoor, L.: Collective spammer detection in evolving multi-relational social networks. In: Proceedings of 21th ACM SIGKDD International Conference on Knowledge Discovery and Data Mining, pp. 1769–1778. ACM (2015)
7. Garey, M.R., Johnson, D.S.: Computers and Intractability; A Guide to the Theory of NP-Completeness. W. H. Freeman & Co., New York (1990)
8. Gjertsen, R.K., Jones, M.T., Plassmann, P.E.: Parallel heuristics for improved, balanced graph colorings. J. Parallel Distrib. Comput. **37**, 171–186 (1996)
9. Gómez, D., Montero, J., Yáñez, J.: A coloring fuzzy graph approach for image classification. Inf. Sci. **176**(24), 3645–3657 (2006)
10. Hastings, W.: Monte Carlo sampling methods using Markov chains and their applications. Biometrika **57**(1), 97–109 (1970)
11. Janis, P., et al.: Device-to-device communication underlaying cellular communications systems. Int. J. Commun. Netw. Syst. Sci. **2**(3), 169–178 (2009). ISSN 1913-3715
12. Jerrum, M.R.: A very simple algorithm for estimating the number of k-colourings of a low-degree graph. Random Struct. Algorithms **7**, 157–165 (1995)
13. Jones, M.T., Plassmann, P.E.: A parallel graph coloring heuristic. SIAM J. Sci. Comput. **14**, 654–669 (1992)
14. Lu, H., Halappanavar, M., Chavarria-Miranda, D., Gebremedhin, A., Kalyanaraman, A.: Balanced coloring for parallel computing applications. In: Proceedings of the IEEE 29th International Parallel and Distributed Processing Symposium, pp. 7–16 (2015)
15. Luby, M.: A simple parallel algorithm for the maximal independent set problem. In: Proceedings of the 17th Annual ACM Symposium on Theory of Computing, pp. 1–10 (1985)

16. Lupaşcu, C.A., Tegolo, D., Bellavia, F., Valenti, C.: Semi-automatic registration of retinal images based on line matching approach. In: Proceedings of the 26th IEEE International Symposium on Computer-Based Medical Systems, pp. 453–456. IEEE (2013)
17. Meer, P.: Stochastic image pyramids. CVGIP **45**(3), 269–294 (1989)
18. Meyer, W.: Equitable coloring. Am. Math. Mon. **80**, 920–922 (1973)
19. Mosa, M.A., Hamouda, A., Marei, M.: Graph coloring and aco based summarization for social networks. Expert. Syst. Appl. **74**, 115–126 (2017)
20. Robert, J., Gjertsen, K., Jones, M.T., Plassmann, P.: Parallel heuristics for improved, balanced graph colorings. J. Parallel Distrib. Comput. **37**, 171–186 (1996)
21. Tantipathananandh, C., Berger-Wolf, T., Kempe, D.: A framework for community identification in dynamic social networks. In: Proceedings of the 13th ACM SIGKDD International Conference on Knowledge Discovery and Data Mining, pp. 717–726. ACM (2007)
22. Tsolkas, D., Liotou, E., Passas, N., Merakos, L.: A graph-coloring secondary resource allocation for D2D communications in LTE networks. In: IEEE 17th International Workshop on Computer Aided Modeling and Design (CAMAD), pp. 56–60. IEEE (2012)
23. Voss, J.: An Introduction to Statistical Computing: A Simulation-based Approach. Wiley, Chichester (2013)

An Attributed Graph Embedding Method Using the Tree-Index Algorithm

Yuhang Jiao[✉], Yueting Yang, Lixin Cui, and Lu Bai

Central University of Finance and Economics, Beijing, China
jiaoyuhang@email.cufe.edu.cn, jiaoyh52@foxmail.com

Abstract. In this paper, we propose an embedding method for attributed graphs. For an attributed graph, we commence by using a tree-index method with the objective of strengthening the vertex labels. For each iteration of the tree-index method, we compute a probability distribution in terms of the frequency of the strengthened labels. With each probability distribution, we compute a Shannon entropy to measure the uncertainty of the strengthened labels. For an attributed graph, with the required Shannon entropies of different TI iterations to hand, we compute an entropy trace vector by measuring how the entropies vary with the increasing TI iterations (i.e., we embed the attributed graph into a vectorial space). We explore our method on several standard graph datasets abstracted from bioinformatics databases. The experimental results demonstrate the effectiveness and efficiency of our method. Our method can easily outperform state of the art methods in terms of the classification accuracy.

1 Introduction

In pattern recognition, graph based representations are powerful tools for structural analysis. Unfortunately, most of the standard pattern recognition and machine learning algorithms are developed for vectors, and are not available for graphs. One way to overcome this problem is to embed the graph data into a vector space, and then deploy vectorial methods. Specifically, in the embedding space, similar graph structures are expected to be close while dissimilar ones far apart.

In order to embed graphs into a vector space, Riesen and Bunke [11] have proposed a dissimilarity embedding method for graphs. For a sample graph, they compute the edit distance from the graph to a number of prototype graphs to give a vectorial description of the graph in the embedding space. Similar to the work of Riesen and Bunke, Bai and Hancock [2] have developed a new dissimilarity embedding method by computing the Jensen-Shannon divergence between a sample graph and a number of prototype graphs. For a sample graph and a prototype graph, the Jensen-Shannon divergence is computed by measuring the entropy difference between the individual graph entropies and a composite

---

Y. Jiao and Y. Yang—Equally contributed.

© Springer Nature Switzerland AG 2019
D. Conte et al. (Eds.): GbRPR 2019, LNCS 11510, pp. 172–182, 2019.
https://doi.org/10.1007/978-3-030-20081-7_17

entropy of a composite structure formed by the graphs. Wilson et al. [13] have proposed an embedding method by representing a graph structure using permutation invariant polynomials that are computed from the spectrum matrix based on algebraic graph theory. Ren et al. [10] have proposed an embedding method by computing permutation invariant features of a graph via the Ihara zeta function. Here, each feature represents the number of a class of main cycles. All these methods bridge the gap between the powerful graph based representation and the algorithms available for the vector based representation. Unfortunately, these methods tend to request burdensome computation for graphs of large sizes (e.g., a graph having thousands of vertices).

To overcome the shortcoming, a family of graph entropy measure methods have been developed. Examples include (1) the approximated von Neumann entropy developed by Han et al. [8], and (2) the Shannon entropy from the information functionals developed by Dehmer et al. [5,6]. Since the computational complexities of these methods are only in quadratic or cubic number of graph vertices, both the methods can be efficiently computed. Unfortunately, the graph entropy measure methods only provide us an one dimensional feature for graphs, and thus cannot reflect interior graph topology information. To overcome this ineffectiveness, Bai and Hancock [1] have developed a novel framework of measuring depth-based complexity traces for graphs. For a graph, they commence by identifying a centroid vertex which has the minimum shortest path length variance to the remaining vertices. For the graph, a family of centroid expansion subgraphs is derived from the centroid vertex. The depth-based complexity trace of the graph is computed by measuring how the entropies of the centroid expansion subgraphs vary with the increasing size of the subgraphs. The complexity trace of a graph can not only be efficiently computed but also provide us a high dimensional entropy based features. Unfortunately, all these existing methods are restricted on unattributed graphs.

To overcome the shortcomings of existing methods, in this paper we aim to propose a novel graph embedding method for attributed graphs. For an attributed graph, we commence by performing a tree-index (TI) label strengthening method (i.e., the TI method defined in [4]) for the purpose of strengthening the vertex labels. For each iteration h of the TI method, we compute a probability distribution in terms of the frequency of the strengthened labels. With the probability distribution for each iteration h, we compute a label Shannon entropy to measure the uncertainty of the strengthened labels. For an attributed graph, we thus compute a new entropy trace by measuring how the label Shannon entropies vary with the increasing TI iteration h (i.e., we embed the attributed graph into a vectorial space). Since the computational complexity of the TI method on an attributed graph is only linear in the number of edges or quadratic in the number of vertices, our new embedding method not only provides us a high dimensional entropy features for the graph but also can be efficiently computed. We explore our method on standard graph datasets abstracted from some bioinformatics databases. The experimental results demonstrate the effectiveness and efficiency of our method. Our method can easily outperform state of the art methods in terms of the classification accuracy.

Section 2 gives the concept of a tree-index based vertex label strengthening algorithm. Moreover, a label Shannon entropy is also defined. Section 3 gives the definition of the new embedding method for attributed graphs. Section 4 provides our experimental evaluation. Finally, Sect. 5 concludes our work.

2   A Vertex Label Strengthening Method

In this section, we describe how to use a tree-index method to strengthen the vertex labels for an attributed graph. We commence by reviewing the definition of a tree-index vertex label strengthening method described in [4]. Then we show how to compute a label Shannon entropy for the probability distribution over the strengthened labels.

2.1   A Tree-Index Based Vertex Label Strengthening Method

In this subsection, we review the concept of a TI method that has been introduced in [4] for strengthening the vertex label of a graph. Assume an attributed graph $G(V, E)$ with vertex set V and edge set E, the discrete label of a vertex $v \in V$ is denoted as $f(v)$. Using the TI method, the new strengthened label for vertex v at the iteration h is defined as

$$TI_h(v) = \begin{cases} f(v) & \text{if } h = 0, \\ \cup_u \{TI_{h-1}(u)\} & \text{otherwise.} \end{cases} \qquad (1)$$

where each vertex $u \in V$ is adjacent to vertex v. At each iteration h, the TI method takes the union of neighbouring vertex label lists as a new label list for v from the last iteration $h - 1$ (the initial step is identical to listing). This procedure creates an iteratively deeper list corresponding to a subtree rooted at v of height h.

Unfortunately, the above method may lead to a rapid explosion of the strengthened label length. Furthermore, taking the union of the neighbouring label lists does ignore the original vertex label information, since the union dose not take any information of the original label for the rooted vertex. One way to overcome this problem is to strengthen the label of a vertex by taking the union of both its label and its neighbouring vertex labels, at each iteration h. Moreover, we use the Hash function for the purpose of compressing the augmented label into a new short label. For the graph G, the neighbourhood of a vertex $v \in V$ is $\mathcal{N}(v) = \{u|(v, u) \in E\}$. For G and each vertex v, the pseudocode of the TI algorithm associated with a Hash function at iteration h is shown in Algorithm 1. Note that, for step 4 of Algorithm 1, we use the same vertex label function F for any graph. This guarantees that all the identical labels of different graphs are mapped into the same index.

Algorithm 1. Strengthen the vertex label using TI method

1: Initialization.
   - Input an attributed graph $G(V, E)$.
   - Set $h=0$. Initialize the vertex labels. For a vertex v of G, assign the original label $feature(v)$ as the initial label $\mathcal{L}_h(v)$.

2: Sort the labels of neighbourhoods for each vertex.

   - For each vertex v of G, sort the labels of its neighbourhood $\mathcal{N}(v)$ in ascending order as $\mathcal{L}_{\mathcal{N}}^h(v) = \{\mathcal{L}_h(u)|u \in \mathcal{N}(v)\}$.

3: Update the label for each vertex.

   - Set $h=h+1$. For each vertex v of G, assign a new label as $\mathcal{L}_h(v) = \{\mathcal{L}_{h-1}(v), \mathcal{L}_{\mathcal{N}}^{h-1}(v)\}$.

4: Compress the vertex label into a new short label.

   - Using the a vertex label function (i.e. the Hash function) $F : \mathcal{L} \rightarrow \Sigma$, compress the label $\mathcal{L}_h(v)$ into a new short label index for each vertex v of G as

$$\mathcal{L}_h(v) = F(\mathcal{L}_h(v)). \tag{2}$$

5: Check h.

   - Check h. Repeat steps 2, 3 and 4 until the iteration h achieves an expected value.

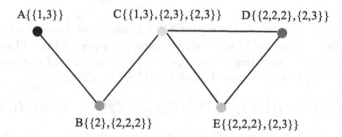

2.2   A Label Shannon Entropy

In this subsection, we compute a Shannon entropy associated with the label probability distribution for an attributed graph [3]. This entropy measures the ambiguity of the subtrees corresponded by the particular strengthened vertex labels. Specifically, let $L = \{l_1, \ldots, l_i, \ldots, l_I\}$ be a label set that contains all possible vertex labels for different graphs, including both the original and strengthened vertex labels. Given an attributed graph $G(V, E)$ and its compressed strengthening label $\mathcal{L}_h(v)$ defined in Eq. (2) for each vertex $v \in V$ at iteration h, we commence by computing the frequency of each particular label l_i contained in

$G(V, E)$, i.e. $n_G^h(l_i)$ for iteration h. The probability $p_G^h(l_i)$ of a label l_i for $G(V, E)$ at iteration h is

$$p_G^h(l_i) = \frac{n_G^h(l_i)}{\sum_{i=1}^{I} n_G^h(l_i)}. \tag{3}$$

With the probability distribution $P_G^h = \{p_G^h(l_1), \ldots, p_G^h(l_i), \ldots, p_G^h(l_I)\}$ of $G(V, E)\}$ to hand, we compute the Shannon label entropy H_S^L for $G(V, E)$ at iteration h as

$$H_S^L(G) = H_S^L(P_G^h) = \sum_{i=1}^{I} p_G^h(l_i) \log p_G^h(l_i). \tag{4}$$

3   The Embedding Method for Attributed Graphs

In this section, we commence by defining a new embedding method for attributed graphs by using the TI method introduced in Sect. 2. Furthermore, we also give the computational complexity analysis of our new embedding method.

3.1   The Attributed Graph Embedding Method Through the TI Algorithm

In this subsection, we investigate how to embed an attributed graph into a vector by measuring the different label Shannon entropies from the different TI iteration.

Definition: For an attributed graph $G(V, E)$, we commence by strengthening the vertex labels using the TI method. For each iteration h of the TI method, let P_G^h is the probability distribution in terms of the frequency of the strengthened labels. The entropy trace vector ET_G for $G(V, E)$ is defined as

$$ET_G = \{H_S^L(P_G^0), H_S^L(P_G^1), \ldots, H_S^L(P_G^2), \ldots, H_S^L(P_G^h), \ldots, H_S^L(P_G^H)\}^T, \tag{5}$$

where H is the largest number of the TI iteration h, and $H_S^L(\cdot)$ is the label Shannon entropy associated with the probability distribution P_G^h defined in Eq. (4). Note that, the label Shannon entropy $H_S^L(P_G^0)$ is computed based on the original vertex label.

Clearly, the dimension of the embedding vector ET_G from the proposed method relates to the largest TI iteration number H, i.e., the dimension equals to $H + 1$. The embedding vector from Eq. (5) provides a high dimensional vectorial representation for an attributed graph.

3.2   The Computational Complexity Analysis

In this subsection, we give the computational analysis of the new attributed graph embedding method. For N graphs (each graph has n vertices) and their label set L, computing the embedding vectors from Eq. (5) requires time complexity $O(HN^2n^2)$. This is because computing the compressed strengthened labels for a graph at each iteration h ($0 \leq h \leq H$) needs to visit all the n^2 entries of the adjacency matrix, and thus requires time complexity $O(Hn^2)$ for all the H iterations. Computing the probability distribution for a graph requires time complexity $O(HNn^2)$ (for the worst case, i.e. each vertex label for the N graphs at all the H iterations are all different and there thus are NHn different labels in L), because it needs to visit all the HNn entries in L for the n vertices. Computing the label Shannon entropy for each graph requires time complexity $O(HNn)$. As a result, the complete time complexity is $O(HN^2n^2)$. This verifies that the proposed embedding method for attributed graphs can be computed in a polynomial time.

4   Experimental Evaluation

We empirically evaluate the performance of the proposed embedding method for attributed graphs. Our experimental evaluation consists of two parts. First, we test the embedding method on classification problem using standard graph datasets. These datasets are abstracted from bioinformatics. Second, we evaluate the stability of the method.

Table 1. Information of the graph based bioinformatics datasets

Datasets	MUTAG	NCI1	NCI109	ENZYMES	D&D	CATH1	CATH2
Max # vertices	28	111	111	126	5748	568	568
Min # vertices	10	3	4	2	30	44	143
Mean # vertices	17.93	29.87	29.68	32.63	284.32	205.70	308.03
# graphs	188	4110	4127	600	1178	712	190
# classes	2	2	2	6	2	2	2

4.1   Datasets

We demonstrate the performance of our new embedding method on seven standard graph datasets from bioinformatics databases [1]. These datasets include: the MUTAG, NCI1, NCI109, ENZYMES, D&D, CATH1 and CATH2 datasets. More details of these datasets are shown in Table 1.

MUTAG: The MUTAG dataset consists of graphs representing 188 chemical compounds, and aims to predict whether each compound possesses mutagenicity. The maximum, minimum and average number of vertices are 28, 10 and 17.93 respectively. As the vertices and edges of each compound are labeled with a real number, we transform these graphs into unweighted graphs.

NCI1 and NCI109: The NCI1 and NCI109 datasets consist of graphs representing two balanced subsets of datasets of chemical compounds screened for activity against non-small cell lung cancer and ovarian cancer cell lines respectively. There are 4110 and 4127 graph based structures in NCI1 and NCI109 respectively. The maximum, minimum and average number of vertices in NCI1 and NCI109 are 111, 3 and 29.87, and 111, 4 and 29.68 respectively.

ENZYMES: The ENZYMES dataset consists of graphs representing protein tertiary structures consisting of 600 enzymes from the BRENDA enzyme database. In this case the task is to correctly assign each enzyme to one of the 6 EC top-level classes. The maximum, minimum and average number of vertices are 126, 2 and 32.63 respectively.

D&D: The D&D dataset contains 1178 protein structures. Each protein is represented by a graph, in which the vertices are amino acids and two vertices are connected by an edge if they are less than 6 Angstroms apart. The prediction task is to classify the protein structures into enzymes and non-enzymes. The maximum, minimum and average number of vertices are 5748, 30 and 284.32 respectively.

CATH1 and CATH2: The CATH1 dataset consists of proteins in the same class (i.e. Mixed Alpha-Beta), but the proteins have different architectures (i.e. Alpha-Beta Barrel vs. 2-layer Sandwich). CATH2 contains proteins in the same class (i.e. Mixed Alpha-Beta), architecture (i.e. Alpha-Beta Barrel), and topology (i.e. TIM Barrel), but in different homology classes (i.e. Aldolase vs. Glycosidases). The CATH2 dataset is harder to classify, since the proteins in the same topology class are structurally similar. The protein graphs are 10 times larger in size than chemical compounds, with 200−300 vertices. There are 712 and 190 test graphs in the CATH1 and CATH2 datasets.

4.2 Experiments on Graph Datasets

Experimental Setup: We evaluate the performance of our attributed graph embedding method (AGEM) on graph classification problems. We also compare our method with alternative state of the art graph based learning methods. These methods include (1) the von-Neumann thermodynamic depth complexity (VNTD) [7,8], (2) the von-Neumann graph entropy (VNGE) [8], (3) the Shannon entropies using the information functionals f^{V_1} (FV1) and f^{P_1} (FP1) [5], (4) the coefficients from the Ihara zeta function for graphs (CIZF) [10], and (5) the hybrid reproducing kernel (HRK) [14].

For all methods, we calculate the vectors or characterization values of graphs as features. We then perform 10-fold cross-validation using the Support Vector Machine Classification (SVM) associated with the Sequential Minimal Optimization (SMO) [12] and the Pearson VII universal kernel (PUK) [9] to evaluate the performance of our method and the alternative methods. We use nine folds for training and one fold for testing. For each method, we repeat the experiments 10 times. All parameters of the SMO-SVMs were optimized for each method on different datasets on a Weka workbench. We report the average classification accuracies of each method and the Runtime in Tables 2 and 3. The runtime is measured under Matlab R2015a running on a 2.5 GHz Intel 2-Core processor (i.e. i5-3210m).

Note that, for our method we vary set the largest iteration for the TI method as 10. The reason for this is that the strengthened labels of all the vertices over all the graphs at iteration $h = 10$ are nearly all different. In other word, after $h = 10$ the probability distributions in terms of the frequency of the vertex labels are nearly the same.

Experimental Results: In terms of the classification accuracies, our AGEM embedding method outperforms all the alternative methods, excluding the FP1 entropy measure method on the MUTAG dataset. The reasons of the effectiveness of our AGEM embedding method are explained as follows.

- **Compared to** the VNTD, VNGE, FV1, FP1 and HRK complexity or entropy methods, our AGEM method can provide us a high dimensional vectorial representation for a graph. By contrast, the VNTD, VNGE, FV1 and FP1 methods only represent a graph in an one dimensional space in terms of the complexity or the entropy feature. Furthermore, our AGEM can also capture the attributed information residing on the vertices through the TI vertex label strengthening method. By contrast, the VNTD, VNGE, FV1, FP1 and HRK methods cannot accommodate the vertex label information.
- **Compared to** the CIZF method, both our AGEM method and the CIZF method can provide us a high dimensional representation for a graph. The CIZF method represents a graph in terms of a set of polynomial coefficients (i.e., the number of different main cycles) via the Ihara zeta function. However, like the VNTD, VNGE, FV1, FP1 and HRK methods, the CIZF method is also restricted on unattributed graphs. By contrast, our AGEM method can accommodate the attributed graphs.
- **Through Tables** 2 and 3, we also observe that the CIZF, FV1 and FP1 methods may generate infinite feature values for graphs of large sizes, since the average sizes of the graphs in the D&D, CATH1 and CATH2 datasets are obviously larger than those in other datasets. This indicates that our AGEM method can also accommodate graphs of large sizes.

In terms of the runtime, our AGEM method is not the fastest method, but our AGEM can finish the computation in a polynomial time on any dataset. Key to the efficiency is that the computational complexity of the TI method on an attributed graph is only linear in the number of edges or quadratic in the number of vertices. By contrast, some methods (i.e., the VNTD and CIZF methods) cannot finish the computation on some datasets, which contain graphs of large sizes, in one day.

4.3 Stability Evaluation

In this subsection, we investigate the stability of our attributed graph embedding method AGEM. We randomly generate two seed graphs. We then apply random

Table 2. Classification accuracy (in %) on different datasets.

Datasets	MUTAG	NCI1	NCI109	ENZYMES	D&D	CATH1	CATH2
AGEM	82.44	**65.79**	**66.00**	**33.50**	**76.31**	**99.01**	**77.36**
VNTD	83.51	–	–	30.50	–	–	–
VNGE	85.10	62.21	62.15	22.33	75.38	98.59	75.78
FV1	84.57	62.04	62.15	24.17	–	–	–
FP1	**85.63**	62.09	62.37	23.33	–	–	–
CIZF	80.85	60.05	62.79	32.00	–	–	–
HRK	84.46	64.86	65.72	24.38	75.36	94.75	71.15

−: can not be finished in one day or the feature values are infinite.

Table 3. Runtime in seconds on different datasets.

Datasets	MUTAG	NCI1	NCI109	ENZYMES	D&D	CATH1	CATH2
AGEM	4″	2′20″	2′20″	23″	7′10″	2′50″	1′10″
VNTD	17′43″	>1 day	>1 day	4 h 30′	>1 day	>1 day	>1 day
VNGE	1″	1″	1″	1″	1″	1″	1″
FP1	2	8″	8″	1″	–	–	–
FP1	2	8″	8″	1″	–	–	–
CIZF	2″	37″	37″	5″	>1 day	–	–
HRK	3″	3′10″	3′10″	3″	2′30″	18″	4″

−: the feature values are infinite.

edit operations on the seed graphs to simulate the effects of noise. The edit operations are vertex deletion and edge deletion, respectively. For each seed graph, we randomly delete a predetermined fraction of vertices or edges to obtain noise corrupted variants. The feature distance between an original seed graph G_o and its noise corrupted counterpart G_n is defined as their Euclidean distance, defined as

$$d_{G_o,G_n} = \sqrt{(ET_{G_o} - ET_{G_n})^T (ET_{G_o} - ET_{G_n})}. \tag{6}$$

We show the results in Figs. 1 and 2. Figures 1 and 2 show the effects of vertex and edge deletion respectively. The x-axis represents 1% to 35% of vertices or edges are deleted, and the y-axis shows the Euclidean distance d_{G_o,G_n} between the original seed graph G_o and its noise corrupted counterpart G_n. From the experimental results, we observe that there is an approximate linear relationship in each case. This implies that the proposed method possesses ability to distinguish graphs under controlled structural-error.

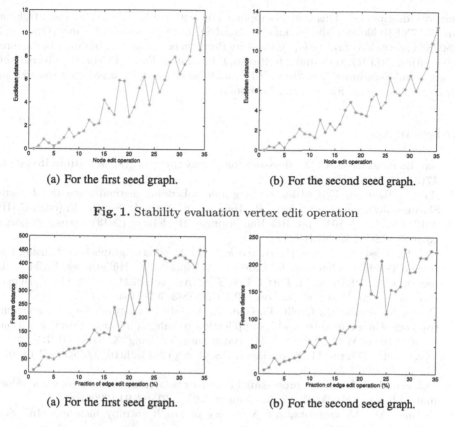

(a) For the first seed graph.   (b) For the second seed graph.

Fig. 1. Stability evaluation vertex edit operation

(a) For the first seed graph.   (b) For the second seed graph.

Fig. 2. Stability evaluation edge edit operation

5 Conclusion

In this paper, we have defined a new embedding method for attributed graphs. Our embedding method is based on a tree-index (TI) label strengthening algorithm on attributed graphs. We compute a label Shannon entropy using the probability distribution associated with the frequency of strengthened labels at each TI iteration h. For an attributed graph, we embed the graph into a vector by measuring how the label Shannon entropies vary with the increasing TI iterations. Comparing most state of the art methods, our method can not only provide us a high dimensional entropy features for graphs but also can be efficiently computed. Moreover, our method also overcomes the restriction on unattributed graphs that arises in the existing graph embedding methods. We explore our method on several standard graph datasets. We demonstrate the effectiveness and efficiency of our method.

Our further work is to extend the tree-index based algorithm used in this paper to hypergraphs, and then define a new embedding method for attributed hypergraphs.

Acknowledgments. This work is supported by National Key R&D Program of China (No. 2017YFB1400700), the National Natural Science Foundation of China (Grant no. 61602535 and 61503422), the Open Projects Program of National Laboratory of Pattern Recognition (NLPR), the Graduate Research Innovation Fund of Central University of Finance and Economics (No. 20181Y019), and the program for innovation research in Central University of Finance and Economics.

References

1. Bai, L., Hancock, E.R.: Depth-based complexity traces of graphs. Pattern Recognit. **47**(3), 1172–1186 (2014)
2. Bai, L., Hancock, E.R., Han, L.: A graph embedding method using the Jensen-Shannon divergence. In: Wilson, R., Hancock, E., Bors, A., Smith, W. (eds.) CAIP 2013. LNCS, vol. 8047, pp. 102–109. Springer, Heidelberg (2013). https://doi.org/10.1007/978-3-642-40261-6_12
3. Bai, L., Rossi, L., Bunke, H., Hancock, E.R.: Attributed graph kernels using the Jensen-Tsallis q-differences. In: Calders, T., Esposito, F., Hüllermeier, E., Meo, R. (eds.) ECML PKDD 2014, Part I. LNCS (LNAI), vol. 8724, pp. 99–114. Springer, Heidelberg (2014). https://doi.org/10.1007/978-3-662-44848-9_7
4. Dahm, N., Bunke, H., Caelli, T., Gao, Y.: A unified framework for strengthening topological node features and its application to subgraph isomorphism detection. In: Kropatsch, W.G., Artner, N.M., Haxhimusa, Y., Jiang, X. (eds.) GbRPR 2013. LNCS, vol. 7877, pp. 11–20. Springer, Heidelberg (2013). https://doi.org/10.1007/978-3-642-38221-5_2
5. Dehmer, M.: Information processing in complex networks: graph entropy and information functionals. Appl. Math. Comput. **201**(1–2), 82–94 (2008)
6. Dehmer, M., Mowshowitz, A.: A history of graph entropy measures. Inf. Sci. **181**(1), 57–78 (2011)
7. Escolano, F., Bonev, B., Hancock, E.R.: Heat flow-thermodynamic depth complexity in directed networks. In: Gimel'farb, G., et al. (eds.) SSPR /SPR 2012. LNCS, vol. 7626, pp. 190–198. Springer, Heidelberg (2012). https://doi.org/10.1007/978-3-642-34166-3_21
8. Han, L., Escolano, F., Hancock, E.R., Wilson, R.C.: Graph characterizations from von Neumann entropy. Pattern Recognit. Lett. **33**(15), 1958–1967 (2012)
9. Maimon, O., Rokach, L. (eds.): Data Mining and Knowledge Discovery Handbook, 2nd edn. Springer, Boston (2010). https://doi.org/10.1007/978-0-387-09823-4
10. Ren, P., Wilson, R.C., Hancock, E.R.: Graph characterization via Ihara coefficients. IEEE Trans. Neural Netw. **22**(2), 233–245 (2011)
11. Riesen, K., Bunke, H.: Reducing the dimensionality of dissimilarity space embedding graph kernels. Eng. Appl. Artif. Intell. **22**(1), 48–56 (2009)
12. Vishwanathan, S.V.N., Sun, Z., Ampornpunt, N., Varma, M.: Multiple kernel learning and the SMO algorithm. In: NIPS, pp. 2361–2369 (2010)
13. Wilson, R.C., Hancock, E.R., Luo, B.: Pattern vectors from algebraic graph theory. IEEE Trans. Pattern Anal. Mach. Intell. **27**(7), 1112–1124 (2005)
14. Xu, L., Jiang, X., Bai, L., Xiao, J., Luo, B.: A hybrid reproducing graph kernel based on information entropy. Pattern Recognit. **73**, 89–98 (2018)

A Graph-Theoretic Framework
for Summarizing First-Person Videos

Abhimanyu Sahu and Ananda S. Chowdhury[✉]

Department of Electronics and Telecommunication Engineering,
Jadavpur University, Kolkata 700032, India
abhimanyusahu009@gmail.com, as.chowdhury@jadavpuruniversity.in

Abstract. First-person video summarization has emerged as an important problem in the areas of computer vision and multimedia communities. In this paper, we present a graph-theoretic framework for summarizing first-person (egocentric) videos at frame level. We first develop a new way of characterizing egocentric video frames by building a center-surround model based on spectral measures of dissimilarity between two graphs representing the center and the surrounding regions in a frame. The frames in a video are next represented by a weighted graph (video similarity graph) in the feature space constituting center-surround differences in entropy and optic flow values along with PHOG (Pyramidal HOG) features. The frames are finally clustered using a MST based approach with a new measure of inadmissibility for edges based on neighbourhood analysis. Frames closest to the centroid of each cluster are used to build the summary. Experimental comparisons on two standard datasets clearly indicate the advantage of our solution.

Keywords: First-person video · Center-surround model ·
Spectral graph dissimilarity · Video similarity graph · MST ·
Inadmissible edge

1 Introduction

Easy availability of more and more commercial wearable devices like GoPro, Google glass, Microsoft Sense-Cam and looxcie cameras [1] enables an user to record huge amount of first-person (egocentric) video data. First-person videos are more challenging to process due to variations like *constant head motion-blur*, *illumination change*, *unstable background* and *frequent changes in people and objects* that naturally occur due to constant movement of the user. Hence, many existing methods for video summarization fail to capture proper summary for the first-person videos. In this paper, we present a graph-theoretic framework for summarizing first-person (egocentric) videos at frame level. The two major contributions of this work are:

(1) We develop a center-surround model (CSM) for ego-centric video frames based on spectral measures of dissimilarity between two graphs representing the center and the surrounding regions in a frame.

© Springer Nature Switzerland AG 2019
D. Conte et al. (Eds.): GbRPR 2019, LNCS 11510, pp. 183–193, 2019.
https://doi.org/10.1007/978-3-030-20081-7_18

(2) We propose a new measure for inadmissible edges in a Minimum Spanning Tree (MST) based clustering applied to the video similarity graph (constructed to represent different frames in a video) to obtain the summary.

2   Related Works

We mention here some of the recently reported works in the field of first-person video summarization. Lee *et al.* [2] created a visual summary of egocentric video by focusing on the most important objects from a video using feature and object segmentation. Lu *et al.* [3] extracted objects from an image and used the detected objects to build a story driven egocentric summarization. In contrast, our proposed model captures more interesting activities/events which often appear in the center of the egocentric video frames. Recently, Guo *et al.* [10] introduced the spatial and temporal scoring mechanism at shot level to summarize egocentric video. For generating a video title, Song *et al.* [5] have applied fully unsupervised approaches. For some works on supervised learning based video summarization, please see [4,6,7]. These supervised learning approaches tend to outperform the unsupervised methods. Very recently, deep learning methods [7–9] become increasingly popular for video summarization. In [7,9], the authors have proposed supervised video summarization using some deep architectures with recurrent models such as LSTMs. The authors in [8] have used deep features for video summarization. Finally, the authors in [14] adopted a graph based hierarchical clustering approach for video summarization. In this paper, we present a graph-theoretic approach with a graph-based center-surround model and a modified MST based clustering for summarizing first-person videos. Our solution is completely unsupervised in nature. To the best of our knowledge, graph based complete solution has not been reported for first-person video summarization.

3   Proposed Method

Our solution pipeline consists of the following steps: (A) Graph based center-surround model, (B) Frame based feature extraction, (C) Construction of the video similarity graph, and (D) MST based improved clustering. An overview of this framework is described in form of a block diagram in Fig. 1. A detailed description of each of the components is now provided below.

3.1   Graph Based Center-Surround Model

In a first person video, important objects mostly tend to appear in the central region of the constituent frames [2]. We propose a graph based center-surround model [15] to better discriminate between the center and surrounding regions in each frame of a egocentric video. A frame f of dimension $W \times H$ is divided into a center region c of dimension $aW \times bH$ and a surrounding region s of dimension $(1-a)W \times (1-b)H$, where $(0 < a, b < 1)$. We now show how to compute for

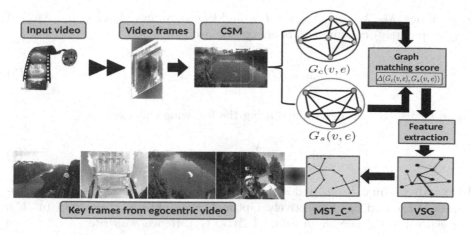

Fig. 1. Schematic of the proposed method (CSM: Center-Surround Model, VSG: Video Similarity Graph, MST_C^*: Modified MST based clustering.)

each frame the optimal set of (a, b) using graph-theory. We represent the center region and the surrounding region by two weighted graphs $G_c(v, e)$ and $G_s(v, e)$ respectively. For $G_c(v, e)$, the pixels within c form the vertices. Similarly, for $G_s(v, e)$, the pixels within s represent the vertices. The edges in both the graphs are established from the 4-connectivity of the pixels (vertices) within that graph. We denote the absolute difference in intensity between two pixels i and j by ΔI_{ij}. So, we write:

$$\Delta I_{ij} = |I_i - I_j| \tag{1}$$

Similarly, the absolute difference in the motion between pixels i and j, represented by difference in the magnitude of the corresponding optical flow vectors $(u(i), v(i))$ and $(u(j), v(j))$ is denoted by ΔM_{ij}. So, we write:

$$\Delta M_{ij} = |\sqrt{u(i)^2 + v(i)^2} - \sqrt{u(j)^2 + v(j)^2}| \tag{2}$$

Spatial affinity between the two pixels i and j is now expressed as the product of $\Delta I(i, j)$ and $\Delta M(i, j)$ and is deemed as the weight w_{ij} of the edge connecting them. Thus, we write:

$$w_{ij} = \Delta I_{ij} \times \Delta M_{ij} \tag{3}$$

We use the spectral measure of graph dissimilarity [16] for G_s and G_c. Let A_{G_c} and A_{G_s} be the weighted adjacency matrices, D_{G_c} and D_{G_s} be the diagonal degree matrices, and L_{G_c} and L_{G_s} be the Laplacian matrices of the graphs G_c and G_s respectively. Then, the Laplacian matrices are given by:

$$L_{G_c} = D_{G_c} - A_{G_c}$$
$$L_{G_s} = D_{G_s} - A_{G_s} \tag{4}$$

We use the similarity matching score $\Delta(G_c(v, e), G_s(v, e))$ between G_c and G_s by computing the difference of the top $k = min(k_1, k_2)$ eigenvalues, where k_1

eigenvalues $(\lambda_{11}, \lambda_{12}, \cdots, \lambda_{1k_1})$ of L_{G_c} and k_2 eigenvalues $(\lambda_{21}, \lambda_{22}, \cdots, \lambda_{2k_2})$ of L_{G_s} (separately) contain 90% of energy. So, we write:

$$\Delta(G_c, G_s) = \sum_{i=1}^{k} (\lambda_{1i} - \lambda_{2i})^2 \tag{5}$$

Here, $k_p, p \in [1, 2]$, is determined using the following equation:

$$\min_j \left(\frac{\sum_{i=1}^{k_p} \lambda_{ji}}{\sum_{i=1}^{n_p} \lambda_{ji}} > 0.9 \right), p \in [1, 2] \tag{6}$$

In the above equation, n_1 and n_2 respectively denote the total number of eigenvalues of L_{G_c} and L_{G_s} respectively. Optimal values of (a, b), denoted by (a^*, b^*), correspond to the maximum value of $\Delta(G_c, G_s)$. Hence, we write:

$$(a^*, b^*) = \underset{a,b \in (0,1)}{\mathrm{argmax}} \{\Delta(G_c, G_s)\} \tag{7}$$

In this way, we find optimal center and surround regions for each frame.

3.2   Frame Based Feature Extraction

In this work, we have used PHOG [17] for feature extraction. We choose PHOG because it represents the local shape and spatial information of the shape. We used three levels of pyramid for building the PHOG features at optimal center region of a video frame. At level 0, the entire frame is considered as one single region and the histogram of edge orientation is calculated for that region. For level 1, the frame is partitioned into four cells and for level 2, the frame is partitioned by into sixteen cells as shown in Fig. 2. Then HOG is determined for 21 (= 1 at level 0 + 4 at level 1 + 16 at level 2) regions. We have also used 12 bins of histogram. Hence, the final PHOG descriptor of the entire center region of a video frame is a vector of size $12 \times \sum_{l=0}^{2} 4^l = 252$. This is illustrated in Fig. 2. Number of levels and bins are chosen to balance accuracy and execution time. For the surrounding region of a video frame, since the information content is somewhat less, we have used two levels of pyramid. Thus, the PHOG descriptor of that region is a vector of size $12 \times \sum_{l=0}^{1} 4^l = 60$.

In addition to PHOG features, we also use differences in the entropy and motion values between the center and the surrounding regions [15]. A video frame is finally represented by a 314-dimensional feature vector (252 elements for PHOG at the center + 60 elements for PHOG at the surrounding + 1 element for center-surround difference in entropy + 1 element for center-surround difference in motion).

3.3   Construction of the Video Similarity Graph

A weighted complete graph, termed as the video similarity graph (VSG) is built in the 314-dimensional feature space for representing all the frames in a video.

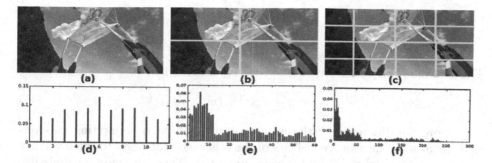

Fig. 2. Extraction of PHOG features from a video frame: partition at different pyramid resolution for different level (top-row): (a) L = 0, (b) L = 1, (c) L = 2 and (bottom row): (d)–(f) concatenation of all the HOG vectors in three pyramid resolutions to obtain the PHOG feature of a frame

Here, each frame acts as a vertex. All the vertices are connected to each-other. The edge weight w_{mn}, between the vertices m and n with respective feature vectors f_m and f_n, is given by:

$$w_{mn} = \sum_{j=1}^{314} |f_{mj} - f_{nj}| \tag{8}$$

We choose the city-block distance for better execution time.

3.4   MST Based Improved Clustering

We apply a MST based clustering in the above VSG with a new measure for inadmissible (inconsistent) edges. After the clustering is complete, a frame nearest to the centroid of each cluster is chosen to build the summary. MST based clustering is chosen because it can detect clusters of any shape (*i.e.*, a cluster need not be regular or convex) and number of clusters need not be known in advance [18]. In MST-based clustering, the inadmissible edges need to be removed successively to get different clusters. To obtain k clusters (where k corresponds to length of the summary), it is necessary to eliminate $k - 1$ inadmissible edges. Performance of MST-based clustering naturally depends on how one determines an inadmissible edge [13]. We propose a new measure of edge inadmissibility by incorporating a neighborhood analysis of the two connected vertices. Since, we also consider weights of the neighborhood edges, the proposed criterion for inadmissibility becomes more robust. Let us consider an edge e_{ab} connecting the vertices a and b. Further, let us assume the degree of a to be m and that of b to be n. Now, we define the following:

$$w_{na} = \sum_{i,i \neq b} w_{ia}/m \tag{9}$$

Similarly,

$$w_{nb} = \sum_{i, i \neq a} w_{ib}/n \qquad (10)$$

The inadmissibility measure for the edge e_{ab} is given by:

$$\delta_{ab} = (w_{na} - w_{ab})^2 + (w_{nb} - w_{ab})^2 \qquad (11)$$

The more the difference of the weight of an edge (w_{ab}) connecting two vertices (a, b) with the average weights of the other incident edges to these vertices (w_{na}, w_{nb}), the higher is the inconsistency (δ_{ab}). Thus, the most inadmissible edge in the MST will have the highest value of δ and so on.

4   Time-Complexity Analysis

We now present the computational complexity of the modified MST-based clustering. Let n be the number of frames in a video. Then, construction of the VSG would take $O(n^2)$ time. This VSG contains n vertices and n^2 edges. Construction of MST from VSG takes $O(n^2 log n^2)$ time. Let the degree of an edge e_{ab} connecting the two vertices a and b in this MST be p and q respectively. Then, δ_{ab} can be computed in $O(p) + O(q)$ time. Let there be r edges in the MST. So, inconsistency measure for all r edges can be computed in $r(O(p) + O(q))$ time. These r edges need to be sorted based on their inconsistency measure, which can be achieved in $O(r^2)$ time. Finally, $(k-1)$ most inadmissible edges to build k clusters can be removed in $(k-1)O(1)$ time. So, the overall complexity is: $(O(n^2) + O(n^2 log n^2) + r(O(p) + O(q)) + O(r^2) + (k-1)O(1)) \approx O(n^2 log n^2)$ (as $p, q, r << n$).

5   Experimental Results

We have used two standard video summarization datasets, namely, SumMe [4] and TvSum50 [5] for experimentation. Detailed information on these datasets are given in Table 1. For performance evaluation, we have used average F-score of all videos [4,5]. All experiments are carried out in MATLAB R2018a environment on a desktop PC with Intel Xeon(R) CPU E5-2690 v4 @ 2.60 GHz, 16 Core and 128 GB of DDR2-memory. Results are compared with two baseline methods and several state-of-the-art approaches [4,5,8,10]. For fair comparisons, we set the threshold of the total duration of key-frame as 15% of the original video length (for both datasets). On average six largest eigenvalues are compared (see Eqs. (5) and (6)) for the graph-based center-surround model.

5.1   Tuning of the Parameters

In this section, we discuss evaluation of the parameters a and b in a way similar to [12]. We demonstrate this with the help of Fig. 3. Here, the surface plot

Table 1. Information about experimental datasets.

Type	Dataset	Video name	Duration (hr:mm:ss)	Frames	Content
Egocentric	SumMe [4]	Base jumping	00:02:38	4729	User videos
		Bike Polo	00:01:42	3064	
		Scuba	00:01:14	2221	
		Val. Downhill	00:02:52	5178	
	TvSum50 [5]	Changing vehicle Tire (VT)	00:09:49	14019	You tube videos
		Getting vehicle unstruck (VU)	00:05:29	9870	
		Parkour (PK)	00:04:34	6580	
		Bee keeping (BK)	00:06:20	11414	

shows variations of the center-surround difference $\Delta(G_c, G_s)$ (along z-axis) with changes in a (along x-axis) and b (along y-axis). The optimal parameter values are those for which the center-surround similarity matching value is maximized. The step-size (of 0.1) chosen for obtaining optimal a and b is based on a trade-off between computational efficiency and accuracy. For a frame from Base jumping dataset, the optimal parameter values are found to be: $a^* = 0.5$ and $b^* = 0.2$. Currently, this process is repeated for each individual frame.

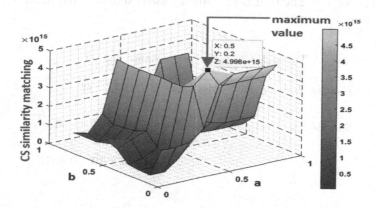

Fig. 3. Parameter estimation strategy a and b

5.2 Ablation Study

We separately show the impacts of (i) graph-based center-surround model (CSM) with adaptive PHOG features (3 levels in the center and 2 levels in the surrounding), denoted by $(PHOG\_CSM)^*$ and (ii) MST based clustering with a new measure for inadmissible edges, denoted by MST_C^* by comparing with a center-surround model without graph and non-adaptive PHOG (2 levels for the entire frame), denoted by $PHOG\_CSM$ and MST-based clustering with simply the weight of an edge as the measure for inadmissibility, denoted by MST_C. Successive improvements in the F-score values are clearly revealed in Table 2.

Table 2. Ablation study based performance analysis.

Dataset	Video name	Approach	F-score
SumMe	Base jumping	$(PHOG\_CSM) + MST\_C$	0.1840
		$(PHOG\_CSM)^* + MST\_C$	0.2331
		$(PHOG\_CSM)^* + MST\_C^*$	**0.2688**

5.3 Cluster Validation

Our proposed measure of inadmissibility is expected to yield better clustering results. We next demonstrate this using two cluster validation measures, namely, Calinski-Harabasz Index (CHI) (the higher, the better) and Davies-Bouldin Index (DBI) (the lower the better) [11]. For this part of the experimentation, we keep the same set of features $(PHOG\_CSM)^*$ and compare K-means, MST and MST_C^*. The results in Table 3, clearly indicate that MST_C^* yields the most superior results.

Table 3. Clustering performance analysis

Dataset	Video name	Methods	$(PHOG\_CSM)^*$	
			CHI ($\uparrow$)	DBI ($\downarrow$)
SumMe	Base jumping	K-means	0.0696×10^3	0.9040
		MST_C	0.1399×10^3	0.7634
		MST_C^*	$\mathbf{0.1595 \times 10^3}$	**0.6575**

5.4 Results on SumMe Dataset

We compare our method with four state-of-the-art video summarization methods, namely, [4,5,8,10] and two baseline methods (K-means clustering and MST clustering) on this dataset. For K-means and HK-means, we set K equal to $l = 15\%$ of the length of the input video. The keyframes closest to the cluster centers are included for the final summary. The results, presented in Table 4, indicate that our approach is almost $19.73 \simeq 20\%$ better than [4,5,10] as well as [8], which uses deep semantic features (using VGG net) on SumMe [4] dataset.

Table 4. Experimental results on SumMe dataset

Video name	Computational methods						
	K-means	Gygli et al. [4]	Song et al. [5]	Otani et al. [8]	Guo et al. [10]	OURS (MST_C)	OURS (MST_C^*)
Base jumping	0.1523	0.1210	0.2990	0.0770	0.2050	0.2331	0.2688
Bike polo	0.1464	0.3560	0.1839	0.2350	0.1910	0.2256	0.2571
Scuba	0.1701	0.1840	0.1390	0.1540	0.2040	0.2061	0.2490
Val. Downhill	0.1522	0.2400	0.2482	0.2580	0.2500	0.2150	0.2587
Mean:	0.1552	<u>0.2252</u>	0.2176	0.1810	0.2125	0.2199	**0.2584**

In Fig. 4, we show the important key frames (key actions) as a part of the summary which match very well with the ground truth scores of an egocentric video entitled "Base jumping" using the proposed method. Our summary captures key actions like *person used head mounted "GoPro" camera for recording, opening the parachute, flying with the parachute* and *the exact moment of landing* (from left to right). These frames are however reported as missed by a recent first-person video summarization method [4].

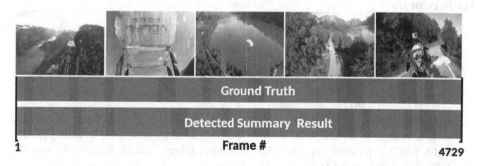

Fig. 4. Comparison of summary obtained using ours method (*bottom bar*) with respect to ground-truth (*top-bar*) where red label indicates important keyframe regions and green label indicates the normal frames for a base jumping video from the SumMe dataset. Frames show important actions. (Color figure online)

5.5 Results on TvSum50 Dataset

For the TvSum50 dataset [5], we compare our method with [5] and two baseline methods (K-means clustering and MST clustering). In Table 5, the results clearly indicate that our approach, with F-score 0.5242 outperforms K-means, MST and [5].

Table 5. Experimental results on TvSum50 dataset

Video name	Computational methods			
	K-means	Song et al. [5]	OURS (MST_C)	OURS (MST_C^*)
Changing vehicle Tire (VT)	0.4247	0.5200	0.5105	0.5197
Getting vehicle unstruck (VU)	0.4063	0.5500	0.5355	0.5591
Parkour (PK)	0.4010	0.4400	0.4965	0.5298
Bee keeping (BK)	0.3946	0.4700	0.4640	0.4880
Mean:	0.4066	0.4950	0.5016	**0.5242**

6 Conclusion

In this paper, we presented a solution for first-person video summarization using a graph-theoretic framework. The two major contributions include a graph-based center-surround model for characterizing ego-centric video and a new measure of edge inadmissibility in MST based clustering. Experimental comparisons clearly show the advantage of our formulation. In future, we plan to experiment with other features like SIFT or SURF as well as deep features (CNN). Another direction of future research will be to explore graph kernels and graph entropies in our model to achieve better accuracy.

References

1. Del Molino, A.G., Tan, C., Lim, J.H., Tan, A.H.: Summarization of egocentric videos: a comprehensive survey. IEEE THMS **47**(1), 65–76 (2017)
2. Lee, Y.J., Ghosh, J., Grauman, K.: Discovering important people and objects for egocentric video summarization. In: CVPR, pp. 1346–1353 (2012)
3. Lu, Z., Grauman, K.: Story-driven summarization for egocentric video. In: CVPR, pp. 2714–2721 (2013)
4. Gygli, M., Grabner, H., Riemenschneider, H., Van Gool, L.: Creating summaries from user videos. In: Fleet, D., Pajdla, T., Schiele, B., Tuytelaars, T. (eds.) ECCV 2014. LNCS, vol. 8695, pp. 505–520. Springer, Cham (2014). https://doi.org/10. 1007/978-3-319-10584-0_33
5. Song, Y., Vallmitjana, J., Stent, A., Jaimes, A.: TVSum: summarizing web videos using titles. In: CVPR, pp. 5179–5187 (2015)
6. Gygli, M., Grabner, H., Van Gool, L.: Video summarization by learning submodular mixtures of objectives. In: CVPR, pp. 3090–3098 (2015)
7. Zhang, K., Chao, W.-L., Sha, F., Grauman, K.: Video summarization with long short-term memory. In: Leibe, B., Matas, J., Sebe, N., Welling, M. (eds.) ECCV 2016. LNCS, vol. 9911, pp. 766–782. Springer, Cham (2016). https://doi.org/10. 1007/978-3-319-46478-7_47
8. Otani, M., Nakashima, Y., Rahtu, E., Heikkilä, J., Yokoya, N.: Video summarization using deep semantic features. In: Lai, S.-H., Lepetit, V., Nishino, K., Sato, Y. (eds.) ACCV 2016. LNCS, vol. 10115, pp. 361–377. Springer, Cham (2017). https://doi.org/10.1007/978-3-319-54193-8_23

9. Mahasseni, B., Lam, M., Todorovic, S.: Unsupervised video summarization with adversarial LSTM networks. In: CVPR, pp. 2982–2991 (2017)
10. Guo, Z., Gao, L., Zhen, X., Zou, F., Shen, F., Zheng, K.: Spatial and temporal scoring for egocentric video summarization. Neurocomputing **208**, 299–308 (2016)
11. Arbelaitz, O., Gurrutxaga, I., Muguerza, J., Pérez, J.M., Perona, I.: An extensive comparative study of cluster validity indices. Pattern Recogn. **46**(1), 243–256 (2013)
12. Almeida, J., Leite, N.J., Torres, R.D.S.: Vison: video summarization for online applications. Pattern Recogn. Lett. **33**(4), 97–409 (2012)
13. Guimarães, S.J.F., Gomes, W.: A static video summarization method based on hierarchical clustering. In: Bloch, I., Cesar, R.M. (eds.) CIARP 2010. LNCS, vol. 6419, pp. 46–54. Springer, Heidelberg (2010). https://doi.org/10.1007/978-3-642-16687-7_11
14. Dos Santos Belo, L., Caetano Jr., C.A., do Patrocínio Jr., Z.K.G., Guimarães, S.J.F.: Summarizing video sequence using a graph-based hierarchical approach. Neurocomputing **173**, 1001–1016 (2016)
15. Sahu, A., Chowdhury, A.S.: Shot level egocentric video co-summarization. In: ICPR, pp. 2887–2892 (2018)
16. Gangapure, V.N., Nanda, S., Chowdhury, A.S., Jiang, X.: Causal video segmentation using superseeds and graph matching. In: Liu, C.-L., Luo, B., Kropatsch, W.G., Cheng, J. (eds.) GbRPR 2015. LNCS, vol. 9069, pp. 282–291. Springer, Cham (2015). https://doi.org/10.1007/978-3-319-18224-7_28
17. Bosch, A., Zisserman, A., Munoz, X.: Representing shape with a spatial pyramid kernel. In: CIVR, pp. 401–408 (2007)
18. Zahn, C.T.: Graph theoretical methods for detecting and describing gestalt clusters. IEEE Trans. Comput. **C–20**, 68–86 (1970)

Network Time Series Analysis Using Transfer Entropy

Ibrahim Caglar[✉] [iD] and Edwin R. Hancock[iD]

University of York, York YO10 5GH, UK
ic656@york.ac.uk

Abstract. The inference of network representations that capture causal relations in time series is a challenging problem. In this paper, we explore the use of information theoretic tools for characterising information flow between time series, and how to infer networks representing time series data. We explore two different approaches. The first uses transfer entropy as a means of characterising information flow and measures network similarity using Jensen-Shannon divergence. The second uses time series correlation and used Kullback-Leibler divergence to compare the distribution of correlations across edges for different networks. We explore how both weighted and unweighted representations derived from these two characterisations perform on real-world time series data. Experiments on time series data for the New York Stock Exchange show that transfer entropy results in better localisation of temporal anomalies in graph time series. Moreover, the method leads to embeddings of network time series that better preserve their temporal order.

Keywords: Kullback-Leibler divergence · Multidimensional scaling · Transfer entropy

1 Introduction

In complex systems, the analysis of causal relationships between dynamic system components remains a challenging problem. Although there are several ways to characterise the similarity between time series, such as cross-correlation, Granger causality and transfer entropy, the inference of causal relationships is prone to noise and error. As a result, many linear and nonlinear approaches to identify causal relationships have been suggested. For instance, Granger published one of the first examples in 1969 [7]. This method provides a means of robustly inferring the causal relationship between time series, and has been widely used in the field of economics for many years. However, this method relies on a linear model and is not easily adapted to non-linear systems.

In the context of information theory, several techniques have been developed to define the relationship between variables using information theory. Mutual information (MI) is a model-free method [10, 14] that indicates how much information we can obtain about one variable from information conveyed by a second

© Springer Nature Switzerland AG 2019
D. Conte et al. (Eds.): GbRPR 2019, LNCS 11510, pp. 194–203, 2019.
https://doi.org/10.1007/978-3-030-20081-7_19

one. Nevertheless, causal relations cannot be identified by MI alone due to its symmetry. In other words, if we attempt to determine if the first variable is affected by the second variable, the relationship between them will always be the same, and we can not assign a cause to either variable. To this end, transfer entropy (TE) was developed by Schreiber [15]. While mutual information is a symmetric measurement, transfer entropy is an asymmetric measurement between two variables, and thus transfer entropy represents directional information transfer. The TE can also be characterised as a time-lagged conditional mutual information [12].

We explained transfer entropy as a means of both edge inference and edge weighting in our previous work [4]. The weighted graph representation is used to compute graph entropy, primarily with the aim of analysing sector graphs and investigating their advantages over non-sectorial representations of the stock market. The sector graph is a structure which distinguishes between stocks of different kinds (e.g. IT, motor industries, food industries, commodities etc.) and represents the within and between sector interactions using the total transfer entropy for edges within and between sectors. We performed PCA on vectorisation of both the sector graphs and the non-sectorial graphs to obtain time series embeddings. We applied these methods to a dataset developed at York that contains sector class labels for the different stock and the closing prices of stock. We showed that the sector graph captures interesting information not conveyed by the non-sectorial representation.

This paper, on the other hand, aims to explore how network representing time series can be constructed using different edge weighting schemes based on information flow between nodes representing different time series. The first of these uses transfer entropy between nodes, together with Jensen-Shannon divergence as a means of comparing different networks. The second uses the Kullback-Leibler divergence between the distribution of correlation coefficients on the edges as a measure of similarity. We also compare the performance when fully connected graphs and graphs with connections inferred from transfer entropy and time series correlation are used. We do not consider sector graphs. The focus of the paper is which graph representation allows the best determination of anomalies in the stock market time series. The edge weighting is based on information theoretic measures derived from the transfer entropy, rather than transfer entropy itself. To analyse the distribution of graphs, we compute a kernel matrix whose elements are computed from either (a) the Jensen Shannon divergence in the case of transfer entropy or (b) the Kulback-Leibler divergences between the distribution of correlation coefficients for the corresponding edges of different networks. This is used to embed the networks into a Euclidean space using MDS. We apply the resulting methods to a new dataset, which is based in econometric measures rather than raw stock closing price. The conclusion from this study is that the new representations provides a more stable characterisation of network evolution, that is less susceptible to noise in the post-2000 trading period, where computerised trading and other factors result in rather volatile stock market behaviour. In other words, we aim to robustly infer the existence of changes in the pattern of causal relationships for a complex system represented by a graph rather than the individual causal relationships. To do this, we explore different strategies for characterising the significance of each causal relation, i.e. each edge, using an evidential weight.

2   Entropic Analysis of Information Flow on Edges

2.1   Preliminaries

Suppose $G(V, E, W)$ is a weighted graph with vertex set V, edge set $E \subseteq V \times V$, and edge weight set W. We use the transfer entropy or cross-correlation to define an edge weight $w_{u,v}$ for the edge $(u, v) \in E$. The weighted adjacency matrix A is defined as follows

$$A(u, v) = \begin{cases} w_{u,v}, & \text{if } w_{u,v} > threshold. \\ 0, & \text{otherwise.} \end{cases} \tag{1}$$

for an undirected graph $w_{u,v} = 1$. We explore the use of both weighted graphs consisting of edges passing the weight threshold and complete weighted graphs (all possible weighted connections are considered).

The degree matrix of graph G is a diagonal matrix D whose elements are given by $D(u, u) = d_u = \sum_{v \in V} A(u, v)$. The normalised Laplacian matrix of the weighted graph G is defined as $\tilde{L} = D^{-1/2} L D^{-1/2}$, where $L = D - A$ is the Laplacian matrix and has elements

$$\tilde{L} = \begin{cases} 1 & \text{if } u = v \text{ and } d_v \neq 0 \\ \frac{-1}{\sqrt{d_u d_v}} & \text{if } (u, v) \in E \\ 0 & otherwise \end{cases} \tag{2}$$

From the eigenvalues of the normalised Laplacian matrix, i.e. $\tilde{\lambda}_i$, $i = 1, ..., |V|$ we can compute the von Neumann entropy. This quantity was originally defined in quantum mechanics and can be expressed in terms of the Shannon entropy associated with the eigenvalues of the density matrix. The normalised Laplacian matrix $\tilde{L}$ can be interpreted as the density matrix of an undirected graph [13], and the von Neumann entropy of the undirected graph can be defined as,

$$H_{VN} = -\sum_{i=1}^{|V|} \frac{\tilde{\lambda}_i}{|V|} \ln \frac{\tilde{\lambda}_i}{|V|} \tag{3}$$

where $|V|$ is the number of nodes in the graph. Han et.al. have shown how to approximate von Neumann entropy for undirected graph in terms of simple degree statistics using the quadratic approximation to the Shannon entropy $x \ln x \approx x(1 - x)$ [8].

$$H_{VN} \approx 1 - \frac{1}{|V|} - \frac{1}{|V|^2} \sum_{(u,v) \in E} \frac{1}{d_u d_v} \tag{4}$$

This allows the efficient calculation for the network entropy in $O(N^2)$ rather than $O(N^3)$ from the normalised Laplacian spectrum.

2.2   Information Flow Between Time Series and Edge Weighting

Here we are interested in weighted graphs where the edges are assigned an information theoretic measure of their significance. To this end we explore two weighting schemes. Each node of the graph is characterised by a time-series and the edge weight captures the significance of the correlation or information transfer between the time series. We explore two ways of capturing this information transfer. The first is based on the transfer entropy between time series. The second is based on the Kulback-Leibler divergence between the node time series correlations.

Time Series Transfer Entropy; To compute transfer entropy, we first require some basic concepts from information theory. Consider the random variable X, following a probability distribution $P(x)$. The Shannon Entropy [16] of the distribution $P(X)$ is defined as $H(X) = -\sum_{x \in X} P(x) \log_2 P(x)$ The base of the logarithm determines the units used for measuring information, and in base 2 the results are given bits [11] if base the is natural the results are given in nits [6]. The joint entropy of the random variables X and Y is defined as $H(X,Y) = -\sum_{x \in X} \sum_{y \in Y} P(x,y) \log_2 P(x,y)$ and the conditional entropy of X given Y [1] is $H(X \mid Y) = -\sum_{x \in X} \sum_{y \in Y} P(x,y) \log_2 P(x \mid y)$

For the case of three random variables X, Y and Z, the Conditional Mutual Information [5,6,9] of X and Y given Z is then defined as, $I(X,Y|Z) = H(X,Z) + H(Y,Z) - H(Z) - H(X,Y,Z)$ in terms of joint entropies of the random variables. It can be re-written as $I(X,Y|Z) = H(X|Z) + H(Y|Z) - H(X,Y|Z)$.

We can now define the Transfer Entropy $T_{u \to v}$ between the time series for nodes u and v. Suppose that X_t^u is the time series data for node u. Transfer entropy is the information transfer from the distribution of random variable X^v to the distribution of random variable X^u, were these variables represent samples from the time series for the two nodes defining an edge. This can be written as a Conditional Mutual Information

$$T_{u \to v} = I(X_{t+1}^u, X_t^v | X_t^u) = H(X_{t+1}^u | X_t^u) - H(X_{t+1}^u | X_t^u, X_t^v)$$

$$= - \sum_{x_t \in X^u, y_t \in X^v} P(x_{t+1}, x_t, y_t) \log_2 \frac{P(x_{t+1}, x_t, y_t) P(x_t)}{P(x_{t+1}, x_t) P(x_t, y_t)} \qquad (5)$$

at different time epochs t and $t + 1$. Here X_t^u and X_t^v are the past states of X^u and X^v respectively, and t is the time index.

While the mutual information is a symmetric measurement between two variables, the transfer entropy is asymmetric measurement between two variables, as the transfer entropy represents the directional information transfer.

We use the transfer entropy to compute the similarity between different graphs using Jensen-Shannon divergence (JSD) [3] and have subsequently use this for embedding. The JSD graph kernel is $k_{JSD}(P(x), Q(x)) = \ln 2 - JSD(P(x), Q(x))$ where $JSD(P(x), Q(x))$ is the Jensen-Shannon divergence between two probability distribution P(x) and Q(x). The JSD between two graphs defined as,

$$S(i,j) = JSD(G_i, G_j) = H(G_i \oplus G_j) - \frac{H(G_i) + H(G_j)}{2}$$

where $H(G_i)$ is the entropy associated with the probability distribution of graph G_i, and $H(G_i \oplus G_j)$ is the entropy associated with the corresponding probability distribution over of the union graph G_x [2]. We have used transfer entropy to compute the JSD for each constituent edge of the individual graphs G_i and G_j, and the union graph $G_i \oplus G_j$, and have summed these to give the total JSD for each pair of graphs.

Edge Weighting via Time Series Cross-Correlation; We compute the Pearson Correlation coefficient between the node time series to compute an edge-weight. For nodes u and v the Pearson coefficient is

$$\rho(u,v) = Cov(X^u, X^v)/Var(X^u)Var(X^v)$$

where $Cov(X^u, X^v)$ is the covariance of the two time series and $Var(X^u)$ and $Var(X^v)$ are their individual variances. The edge weight is given by $w(u,v) = abs(\rho(u,v))$. The cross-correlation is calculated for all pairs of time series and gives a $V \times V$ cross-correlation matrix. We convert the correlations to probabilities in order to compute Kullback-Leibler Divergence (KLD) between graphs.

To compute the similarity between graphs we use the Kullback-Leibler Divergence (KLD). This is an asymmetric measurement of the difference between the discrete probability distributions P and Q defined on the same probability space.

$$KLD(P||Q) = \sum_{x \in X} P(x) \log_2 \frac{P(x)}{Q(x)}$$

If $KLD(P||Q) = 0$, the distributions are equal. It is asymmetric; $KLD(P||Q) \neq KLD(Q||P)$ and always positive; $KLD(P||Q) \geq 0$.

Turing our attention to the representation of similarity of pairs of graphs based on distribution of correlation coefficient, the edge (u, v) the probability is

$$p_{u,v} = \frac{w_{u,v}}{\sum(w_{u,v})} \tag{6}$$

To populate the elements of the edge probability distribution we vectorise the upper triangular (or lower triangular) component of the edge probability matrix. We compute the KLD with the probabilities in the upper triangle for each pair of graphs. For the graph G_i the probability distribution is $P(G_i)$. The resulting KLD for a pair of graphs is not in general symmetrical due to the properties of KLD listed above, and MDS cannot be applied to asymmetric matrices. To overcome this problem we use the symmetrical KLD (j divergence)

$$S(i,j) = KLD(P(G_i)||P(G_j)) + KLD(P(G_j)||P(G_i))$$

2.3   Embedding Using Multidimensional Scaling

Multidimensional scaling (MDS) is a way of visualising similarity that preserves distances. Suppose we have an $N \times N$ distance or similarity matrix S for a sample of N data, then we can apply double centering,

$$K = (I - \frac{J}{N})S(I - \frac{J}{N}) \tag{7}$$

where I is the $N \times N$ identity, $J = ee^T$ is an $N \times N$ matrix of all 1's and e is the all-ones vector $e = (1, 1, ...1)^T$. The eigendecomposition of K is $K = \Phi\Lambda\Phi^T$, where Λ is the $N \times N$ diagonal matrix containing the ordered eigenvalues on the diagonal, and Φ is the $N \times N$ matrix with the ordered eigenvectors as columns. The matrix containing the embedding co-ordinates is

$$X = \sqrt{\Lambda}\Phi^T$$

We take the leading n rows of X to given an embedding into and n-dimensional space. In our experiments $n = 2$.

We have applied MDS to the similarity matrices for the JSD between edge transfer entropies and the KLD between edge correlation coefficents. The result is an embedding of the time series of graphs.

3   Experiments

In this section, we test and compare our different approaches to edge inference and edge weighting on real-world datasets. The Stock Market Network Dataset gives the daily closing prices from 1986 to 2011 for stock on the New York Stock Exchange (NYSE). This dataset has been subject to earlier analysis [17] involving correlation-based and thermodynamic analysis of topological variations of the market network during financial crises [18]. We use the correlation-based and entropy-based networks to represent the structure of the stock market [4]. Here, we have 347 stock that with historical data from 1986 to 2011 [17]. Each stock has approximately 6000-days of unbroken data. We compute the cross-correlation and transfer entropy between the time series for each pair of stock over a time window of 28 days and create an edge between stocks if the correlation coefficient is among the highest 5% of the total cross-correlation coefficients. Transfer entropy based weight values were calculated similarly.

3.1   Graph Entropy

Our aim is to explore which network characterisation allows the cleanest identification of market crises events. To this end we construct representations based on the time evolution of both edge weighted (correlation-based and transfer-entropy-based) and unweighted market networks. Given the network structure at each time epoch, we compute the von Neumann entropy using weighted from the correlation coefficient (Eq. 6).

For a weighted network according to Eq. 3 and unweighted network according to Eq. 4. We also calculate the mean and standard deviation of the entropy. We threshold the entropy and count how many crises correspond to drops in entropy below the threshold without false positives. For each method we vary the threshold to locate the maximum number of crises without false positives.

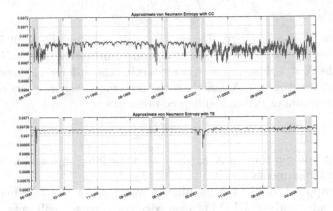

Fig. 1. This figure shows comparison of entropy obtained using the correlation-based unweighted network and transfer-entropy-based unweighted network. Each shaded areas represent different financial crises through time. Red dash-line represent a threshold value to detect crisis points. (Color figure online)

In Fig. 1 we show two plots of entropy. The upper plot shows von Neumann entropy values for the undirected correlation-based network for each time epoch, while the lower plot shows entropy values from the undirected entropy-based network for each time epoch. When there are no market crises, the entropy values generally very slowly. For crises, on the other hand, local entropy anomalies appear. In Figs. 1 and 2 we have shaded areas which represent stock market crashes and financial crises during 1986–2011. Examples include Black Monday, 13th Friday mini-crisis, 1990–1991 recession, 1997 Asian Financial Crisis, and the 27 October mini-crisis, 1998 Russian Financial Crisis, dot-com bubble, 9/11, Chinese Stock bubble, U.S. Bear Market and European Dept Crises, respectively. The red-dash line is a reference line which shows the mean entropy minus 2 standard deviations.

In Fig. 2 the upper plot shows the von Neumann entropy for the correlation-based weighted network for each time epoch, and the lower plot shows the entropy from the entropy-based weighted network. The crises as mentioned earlier refer to the times indicated here as well.

The main features to note from these plots are as follows. First, when we use transfer entropy to either weight or threshold the edges, the crises are more cleanly detected than if we use time series correlation. Second, the weighted transfer entropy network gives the cleanest localisation of crisis entropy anomalies. These conclusions are further supported if we compute the local mean and

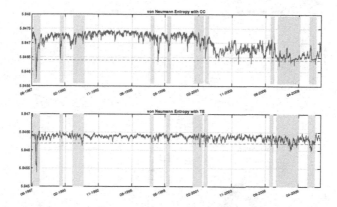

Fig. 2. Comparison between the correlation-based weighted network and transfer-entropy-based weighted networks.

variance of the entropy. Using a time window of 28 days, the variance in local mean is smallest for the unweighted transfer entropy network and largest for the weighted correlation network. These conclusions are supported by the data in Table 1.

Table 1. The first line shows how many crises points cross the red-dashed line without false positive, second line average rolling mean for whole network, and third line is the rolling variance for the whole network

	Unw. CC	Unw. TE	Weighted CC	Weighted TE
Crisis points (M-2std)	0	3	0	**9**
Mean	−3.9420e−09	−5.5790e−11	−6.2928e−08	−5.1781e−09
Variance	−3.1117e−10	**−1.1743e−11**	−1.7885e−09	**−6.1654e−10**

3.2 Network Embedding

For both the KLD and JSD embeddings we visualise the embedded time series in the space spanned by the leading 3 dimensions. The results of the KLD embedding can be seen in the upper two scatter plot in Fig. 3 while the results of the kernel-JSD embedding are shown in the lower two plots. The plots in the left-hand column are obtained from unweighted correlation networks, those in the right-hand column are obtained from unweighted transfer entropy networks.

In order to make the evaluation, we calculated the distance between successive points in the ordinal time sequence. We then determined whether the points were closer to the neighbouring points in the sequence than to out of sequence points. The embeddings contain 5970 points, and in the best-case scenario, 32% and 20% following points were found close to each other, and these graphs generated by TE (Table 2).

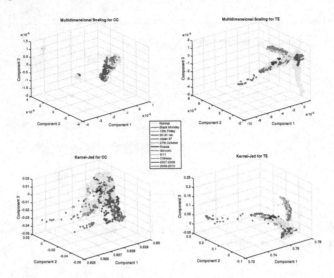

Fig. 3. Two different embedding method produced two different weighed edge network. Different colour represent different crisis times but light-blue represent non-crisis times. (Color figure online)

Table 2. Embedding from correlation-based have minimum average distance between consecutive points however second results show unrelated points closed each other.

	MDS (CC)	MDS (TE)	kernel-JSD (CC)	kernel-JSD (TE)
Average distance	**6.4810e−07**	1.0070e−06	**0.0011**	0.0026
Number of nearest conseq	1596	**1935**	1054	**1216**

4   Conclusion

We have explored how cross-correlation and transfer entropy can be used to construct both unweighted and weighted network representations of time evolving data. We explore how the entropy of these graphs can be used to detect network anomalies. Our conclusion is that transfer entropy outperforms times series cross-correlation. Moreover, the characterisations are more stable if unweighted networks are used, and the two measures are used to threshold edges rather than weighting them.

In the future, we plan to explore whether we can define divergence measures on an edge-by-edge basis rather than in the global manner adopted here. We also intend to explore how transfer entropy can be used to perform analysis on single networks and used to cluster nodes and explore node salience using centrality and related measures.

References

1. Abdul Razak, F., Jensen, H.J.: Quantifying 'causality' in complex systems: understanding transfer entropy. PLoS ONE **9**(6), 1–14 (2014)
2. Bai, L., Hancock, E.R., Ren, P.: Jensen-Shannon graph kernel using information functionals. In: Proceedings - International Conference on Pattern Recognition ICPR, pp. 2877–2880 (2012)
3. Bai, L., Hancock, E.R.: Graph kernels from the jensen-shannon divergence. J. Math. Imaging Vis. **47**(1–2), 60–69 (2013)
4. Caglar, I., Hancock, E.R.: Graph time series analysis using transfer entropy. In: Bai, X., Hancock, E.R., Ho, T.K., Wilson, R.C., Biggio, B., Robles-Kelly, A. (eds.) S+SSPR 2018. LNCS, vol. 11004, pp. 217–226. Springer, Cham (2018). https://doi.org/10.1007/978-3-319-97785-0_21
5. Cover, T.M., Thomas, J.A.: Entropy, Relative Entropy, and Mutual Information. In: Elements of Information Theory, pp. 13–55. Wiley, New York (2005)
6. Frenzel, S., Pompe, B.: Partial mutual information for coupling analysis of multivariate time series. Phys. Rev. Lett. **99**(20), 1–4 (2007)
7. Granger, C.W.J.: Investigating causal relations by econometric models and cross-spectral methods. Econometrica **37**(3), 424 (1969)
8. Han, L., Escolano, F., Hancock, E.R., Wilson, R.C.: Graph characterizations from von Neumann entropy. Pattern Recogn. Lett. **33**(15), 1958–1967 (2012)
9. Hlavackovaschindler, K., Palus, M., Vejmelka, M., Bhattacharya, J.: @Association-Measure@Causality detection based on information-theoretic approaches in time series analysis. Phys. Rep. **441**(1), 1–46 (2007)
10. Kraskov, A., Stögbauer, H., Grassberger, P.: Estimating mutual information. Phys. Rev. E - Stat. Nonlinear Soft Matter Phys. **69**, 66138 (2004)
11. Kwon, O., Yang, J.S.: Information flow between stock indices. EPL (Europhys. Lett.) **82**(6), 68003 (2008)
12. Lee, J., Nemati, S., Silva, I., Edwards, B.A., Butler, J.P., Malhotra, A.: Transfer entropy estimation and directional coupling change detection in biomedical time series. Biomed. Eng. Online **11**(1), 19 (2012)
13. Passerini, F., Severini, S.: The von Neumann entropy of networks. In: Developments in Intelligent Agent Technologies and Multi-Agent Systems, pp. 66–76, December 2008
14. Ross, B.C.: Mutual information between discrete and continuous data sets. PLoS ONE **9**(2), 1–5 (2014)
15. Schreiber, T.: Measuring information transfer. Phys. Rev. Lett. **85**(2), 461–464 (2000)
16. Shannon, C.E.: A mathematical theory of communication. Bell Syst. Tech. J. **27**(3), 379–423 (1948)
17. Silva, F.N., et al.: Modular Dynamics of Financial Market Networks. arXiv e-prints arXiv:1501.05040, January 2015
18. Ye, C., Torsello, A., Wilson, R.C., Hancock, E.R.: Thermodynamics of time evolving networks. In: Liu, C.-L., Luo, B., Kropatsch, W.G., Cheng, J. (eds.) GbRPR 2015. LNCS, vol. 9069, pp. 315–324. Springer, Cham (2015). https://doi.org/10.1007/978-3-319-18224-7_31

Reconstructing Objects from Noisy Images at Low Resolution

Helene Svane[1]([⊠])[iD] and Aasa Feragen[2][iD]

[1] Department of Mathematics, Aarhus University, Aarhus, Denmark
`helenesvane@math.au.dk`
[2] Department of Computer Science (DIKU),
University of Copenhagen, Copenhagen, Denmark
`aasa@di.ku.dk`

Abstract. We study the problem of reconstructing small objects from their low-resolution images, by modelling them as r-regular objects. Previous work shows how the boundary constraints imposed by r-regularity allows bounds on estimation error for noise-free images. In order to utilize this for noisy images, this paper presents a graph-based framework for reconstructing noise-free images from noisy ones. We provide an optimal, but potentially computationally demanding algorithm, as well as a greedy heuristic for reconstructing noise-free images of r-regular objects from images with noise.

Keywords: Object reconstruction · r-regularity

1 Introduction

Whenever new imaging techniques enable us to improve image resolution, we find something smaller scale or further away that we would like to investigate. As a result, the ability to reconstruct objects whose size is on a similar scale as the resolution, is and remains a highly relevant problem, which finds applications in fields as diverse as microscopy and astronomy, see Fig. 1, left. Reconstruction of such small objects is hampered by the fact that all information about the object is contained in just a few pixel intensities. In this paper we assume that the imaged object satisfies r-*regularity*, which reduces the possible complexity of the object and therefore enables inference with bounds on precision (see Sect. 2 for a precise definition of r-regularity).

Previous work by Svane and du Plessis [16] studied ideal images of r-regular objects. By *ideal* images, we mean images of completely black objects placed on completely white backgrounds, taken with a perfect camera so that the intensity of each pixel is exactly equal to the fraction of the pixel that is covered by

This research was supported by Centre for Stochastic Geometry and Advanced Bioimaging, funded by a grant from the Villum Foundation. The authors thank François Lauze and Pawel Winter for valuable discussions.

© Springer Nature Switzerland AG 2019
D. Conte et al. (Eds.): GbRPR 2019, LNCS 11510, pp. 204–214, 2019.
https://doi.org/10.1007/978-3-030-20081-7_20

the original object. In real life, ideal images are rare, maybe even non-existent. Hence we would like to use our knowledge of ideal images to reconstruct r-regular objects from their noisy images. The strength of this approach is that there are relatively strict limitations on which configurations of black, grey and white pixels can occur in an ideal image of an r-regular object. Thus, by considering noisy images as distortions of ideal ones, we aim to use these limitations to deduce the most likely corresponding ideal image. From this idealised image, we apply techniques developed for ideal images to suggest a reconstruction of the original object.

 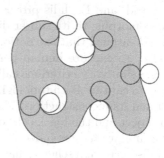

Fig. 1. *Left:* Noisy image of distant galaxies from [12]. Astronomers are interested in knowing the shape of such galaxies: Are they circular, ellipsoid or do they have spiral arms? *Right:* An r-regular set with osculating r-balls shown.

We formulate the idealisation of a noisy image as a graph problem which can be solved using integer linear programming (ILP); this is explained in Sect. 3. As finding an optimal solution using ILP is NP-hard, we also suggest a less computationally demanding greedy algorithm, which makes stepwise locally optimal improvements starting from a trivial initialization. This algorithm generally produces a suboptimal output, but in practice it performs well when aggregated over multiple runs.

2   Related Work

While "object reconstruction" aims to infer any geometric or topological property of the original object, the ultimate goal is to reconstruct the object itself, a task which largely coincides with image segmentation. Modern image segmentation algorithms such as deep convolutional neural networks [8] work as pixel classifiers which cannot possibly return the underlying object itself. At the very best, they return an ideal image of the object. Our proposed algorithm thus should not be viewed as an alternative to these segmentation tools, but rather as a tool to be used together with them, estimating object boundaries from pixel classification output.

2.1   Related Reconstruction Approaches

The task of reconstructing an object from its digital image depends on the image model used and on what kind of properties one wishes the reconstructed set to have. A classical digitisation model is *subset digitisation*, where a spatial object is intersected with a grid and the Voronoi cell centered at a grid point is coloured black if that grid point is inside the object, and white otherwise. Pavlidis [10] and Serra [13] both studied such digitisations and independently introduced r-regular sets in their work on reconstructing a set from the black cells of its subset digitisation. Serra proved that under certain conditions, the homotopy type is preserved under subset digitisation of an r-regular object by a hexagonal grid, and Pavlidis proved that under certain conditions an r-regular set is homeomorphic to its subset digitisation by a square grid. Stelldinger and Köthe [14,15] argued that the notions of homotopy type or homeomorphism are not sufficient to capture human perception of shape similarity and introduced two stronger similarity criterions called *weak* and *strong r-similarity*, with which they aimed to capture both topological properties and physical distance between an object and its reconstruction. They showed that under certain conditions, an r-regular object and the black cells of its subset digitisation are both weakly and strongly r-similar, and extended their results to blurred images.

Other discretisation schemes have also been studied. In [11], Ronse and Tajine study which discretisations are optimal in the sense that given a set X and a discrete set $\mathcal{D}$, which subset(s) of $\mathcal{D}$ minimises the Hausdorff distance to X. In [7], Latecki et al. argue that a realistic digitisation model is obtained by covering the object of interest with a square grid and assigning to each pixel in the grid an intensity which is a monotonic function of the fraction of the pixel that is covered by the object. They show that when applying any threshold to such an image and considering the set of black pixels as the reconstructed set, an r-regular set and its reconstruction is homotopy equivalent under certain conditions, and even homeomorphic when the r-regular set is a manifold with boundary (they conjecture that all r-regular sets are manifolds with boundary - this was later proven by Duarte and Torres in [4]).

Our digitisation model is a special case of the one proposed by Latecki et al. [7], since we will assume each pixel intensity to be exactly the fraction of the pixel covered by the object. However, where Latecki et al. applied a threshold to their resulting image, we have kept the information about the grey pixels in our approach and proposed a reconstruction set with smooth boundary.

2.2   Related Segmentation Algorithms

A number of classical image segmentation algorithms aim to estimate object boundaries from images. For instance, variational segmentation algorithms aim to estimate either the object [9] or its boundary [1,2,6,18] by optimizing a functional which measures the fit between the proposed object (boundary) and the image. In Active Shape Models [3], this is done with the additional information of a statistical object model. In [5], simulated annealing is used to reconstruct the entire object (and therefore, implicitly, its boundary) using a statistical image model based on thermodynamics.

As these methods are based on non-convex optimization, they are not guaranteed to find an optimal solution, and therefore work best when one has some initial idea about what object to find. In contrast, under the assumption of r-regularity, Svane and du Plessis [16] obtain bounds on the error of the reconstructed object boundary, at least in the non-noisy case. We therefore propose estimating the optimal non-noisy image from a noisy one, as explained in Sect. 3. The object boundary can then be reconstructed from the non-noisy image with guarantees.

Note, additionally, that the assumption of r-regularity is a local restriction on the shape of the original object, and not a global assumption as the one that is needed in e.g. the Active Shape Model approach [3].

2.3   Reconstructing r-regular Objects from Ideal Images

The strength of our approach is the assumption of r-regularity of the objects that we are looking for, since this puts restrictions on their possible ideal images. Let us introduce the concept of r-regularity:

Definition 1. *Let $r \in (0, \infty)$. A closed set $X \subseteq \mathbb{R}^n$ is said to be r-regular if, for any point $p \in \partial X$, there exist two balls $B_r(x_b) \subseteq X$ and $B_r(x_w) \subseteq X^c$ of radius r such that $\overline{B_r(x_b)} \cap \overline{B_r(x_w)} = \{p\}$, see Fig. 1, right.*

In [16], Svane and du Plessis studied digital images of r-regular objects constructed in the following way:

Definition 2. *Let $X \subseteq \mathbb{R}^2$ be a subset and $(d\mathbb{Z})^2 \subseteq \mathbb{R}^2$ a lattice. To each lattice square C, we assign an intensity $\lambda \in [0, 1]$ given by*

$$\lambda = \frac{Area(X \cap C)}{d^2} \in [0, 1].$$

The image of X (by $(d\mathbb{Z})^2$) is now the collection of pairs (C, λ) of lattice cubes and their corresponding intensities.

In this paper, we will consider noisy images. In a noisy image, the intensities are distorted, so it will most likely be hard to use intensities for reconstruction. Hence, we may as well restrict ourselves to only consider pixels as being either black, grey or white. We therefore introduce the following:

Definition 3. *Let I be an image. The trinary image J (of I) is the image I where all grey values are set to 0.5. If I is ideal, we call J the trinary ideal image.*

Theorem 1 (Proved in [16]). *Let J be a trinary ideal image of an r-regular object X by a lattice $(d\mathbb{Z})^2$, with $d\sqrt{2} < r$. Then we can construct an object Y from J such that $d_H(\partial X, \partial Y) < d$, where d_H denotes the Hausdorff distance. The running time for this reconstruction algorithm on an $n \times n$ image is $O(n^2)$.*

Empirical results suggest that we can improve the Hausdorff distance between object and reconstruction, but for now it will have to remain a conjecture.

In the process of proving this theorem, the following theorem popped up, and it will be essential later on:

Theorem 2 (Proved in [16]). *In the trinary ideal image of an r-regular object by a lattice $(d\mathbb{Z})^2$ with $d\sqrt{2} < r$, there are at most 562 different configurations of 3×3 pixels. These are the ones shown in Fig. 2, along with their rotations, mirror images and interchanging of black and white colours.*

Furthermore, there are limits on which configurations can be combined with which in such an image.

Fig. 2. Up to rotation, mirroring and interchanging of black and white pixels, these are the only 3×3 configurations of pixels we expect to see in the image of an r-regular object with $d\sqrt{2} < r$

Sketch of proof of Theorem 1: A detailed proof of Theorem 1 is too long for this paper, but we sketch the basic ideas here: Firstly, we prove Theorem 2. This proof is rather long and technical since it involves a lot of cases, but the basic idea is to rule out trinary configurations that do not permit the existence of inner and outer r-balls tangent to some boundary point, as required by r-regularity.

The reconstructed object is then constructed in the following way: Consider two grey pixels sharing a side. Either the pixels are part of some configuration of 2×2 grey pixels, or they are not. If they are, we place an auxiliary point at the midpoint of the 2×2 configuration, otherwise we place an auxiliary point on the midpoint of the common edge between the two grey pixels. After some further manipulation we get a chain of auxiliary points where each two neighbour points in the chain are one grey pixel apart.

We then fit circle arcs through each three consecutive points of auxiliary points and interpolate between the resulting arcs to construct a smooth curve - this is the boundary of our reconstruction. We prove that this curve separates the white pixels from the black ones. Since the original set boundary has the same property, we may put an upper bound on the Hausdorff distance between the original object and our reconstruction. We refer to [16] for details.

3   Noisy Images

Consider a noisy image I of an r-regular object X where $d\sqrt{2} < r$. We want to use our knowledge of trinary ideal images of r-regular objects to find the trinary ideal image of X from the noisy one. Henceforth, the only ideal images we will be working with will be trinary, so we will often omit the word 'trinary' for brevity. We will say that a noisy image I has an underlying (trinary) ideal image J that is not observed.

By Theorem 2 there is a collection $C = \{C_k\}_{k \in K}$ of ideal 3×3 pixel configurations that we may see in the image of an r-regular set when $d\sqrt{2} < r$. We use these to formulate the problem as a graph problem. Let $\mathcal{I}_{i,j}$ denote the 3×3 pixels centered at the (i, j)'th pixel of the noisy image I.

Over each configuration $\mathcal{I}_{i,j}$ in I, the ideal configurations in C are possible configurations in the same position of the underlying ideal image J. These ideal configurations form the vertices of a graph. An ideal configuration $C_k \in C$ sitting over the noisy configuration $\mathcal{I}_{i,j}$ is given a weight $p_k^{i,j}$ measuring the similarity between C_k and $\mathcal{I}_{i,j}$. Two ideal configurations are connected by an edge if they sit over neighbour configurations in I and match on their overlap, see Fig. 3. By 'neighbour configurations' we here mean any two 3×3 configurations sharing 6 pixels.

The problem is now to choose an ideal configuration over each noisy configuration in I, such that the chosen configurations match their neighbours on the overlap, and the sum of their similarity weights are maximised. If this problem is solved, we may piece an ideal image together from the configurations chosen. This image is then optimal in the sense that the sum of its similarity weights is maximal among all ideal images.

The problem can be formulated as an integer linear programming problem in the following way: Let

$$c_k^{(i,j)} = \begin{cases} 1 & \text{if } k'\text{th configuration is chosen at position } (i, j) \\ 0 & \text{otherwise} \end{cases}.$$

The sum to be maximised is then

$$\sum_{i,j,k} c_k^{(i,j)} p_k^{(i,j)}, \quad \text{where} \quad \sum_{k \in K} c_k^{(i,j)} = 1$$

for each (i, j), since we only choose one configuration over each configuration in the noisy image. Stacking $c_k^{(i,j)}$ to a vector c, let A be the adjacency matrix of the graph. Requiring that each chosen configuration is connected to its chosen neighbour configuration may be formulated as $Ac \geq Bc$, where B is a diagonal matrix whose (l, l)'th element is the number b_l of neighbour configurations of configuration l.

Solving these equations is a well-known optimisation problem with a range of software options available, of which we used CPLEX [17].

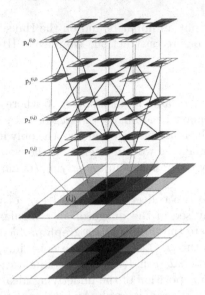

Fig. 3. Over each 3×3 configuration in a noisy image I, we have a set of possible configurations for the underlying image J. These configurations form the vertices of a graph and are given weights quantifying their deviation from the observed configuration. Two configurations are connected by an edge if they sit over neighbour configurations in I and match on their overlap.

4   A Greedy Local Algorithm

As ILPs are generally NP-hard, the running time for the above algorithm quickly increases for large images. Therefore, we also tried another approach: We start with one solution to the graph problem and try to improve it.

Let I be a noisy image with underlying ideal image J. We suggest the greedy algorithm detailed in Algorithm 1, which was implemented in MatLab.

5   Similarity Weights

In Sect. 3 we needed weights $p_k^{i,j}$ on ideal configurations C_k measuring their similarity to an observed noisy configuration $\mathcal{I}_{i,j}$. We propose to construct the weights as follows:

Let p_1 be a pixel from the noisy image and p_2 a trinary pixel from one of the ideal configurations. We measure the distance between p_1 and p_2 with a function $\hat{d}$ given by

$$\hat{d}(p_1, p_2) = \begin{cases} 0 & \text{if } p_2 = 0.5 \text{ and } 0 < p_1 < 1 \\ |p_1 - p_2| & \text{otherwise} \end{cases}$$

We can then define the weight $p_k^{i,j}$ as

$$p_k^{i,j} = \sqrt{\sum_{r,s=1}^{3} \hat{d}(\mathcal{I}_{i,j}(r,s), C_k(r,s))^2},$$

where $\mathcal{I}_{i,j}(r,s)$ denotes the (r,s)'th entry in $\mathcal{I}_{i,j}$.

Algorithm 1. Pseudo-code for the local algorithm

INPUT Noisy image I
$\hat{J}$ = Suggestion for I, initialised to an all white image.
$Weights$ =Matrix of weights of the configurations in current $\hat{J}$
$[XMin, YMin]$ =Position of minimal value of $Weights$
Update configuration at position $[XMin, YMin]$ in $\hat{J}$ to configuration with a black pixel in the middle, and greys around it.
Update $Weights$ to match this new $\hat{J}$
$NbList$ =Positions of horisontal/vertical neighbour configurations of altered ones.
$k = 0$
while k=0 **do**
    TempNbList=NbList;
    Find the entry (i, j) in $TempNbList$ where changing white pixels of corresponding configuration in $\hat{J}$ to greys and centre pixel to black would cause the largest increase in $Weights(i, j)$.
    if Such a configuration exists **then**
        Check if this update of $\hat{J}$ contains any illegal configurations
        if It does not **then**
            Update $\hat{J}$
            Update $Weights$
            Update $NbList$ by adding positions of neighbour configurations
        else
            Remove position (i, j) from $TempNbList$
        end if
    else
        $k = 1$
    end if
end while
if n connected component are expected in the image **then**
    Repeat the above n times
end if

6   Experiments

Results from both the ILP algorithm and the greedy local algorithm are shown in Table 1. The noisy images were obtained from ideal ones by adding Gaussian white noise with mean 0 and variance 0.1. The greedy algorithm was run several times with different starting points (s.p.), giving different outputs. The results using the three most likely starting points are shown in columns 3–5. In column 6 the minimum of 8 outputs of the greedy algorithm with different starting points is shown. Since the minimum image may contain configurations not in the list from Theorem 2, we may not use our reconstruction algorithm on it. Therefore we have used the ILP algorithm on it to remove illegal configurations, see column 7. Finally, in column 8 we have used the ILP algorithm directly on the noisy image. The yellow lines in the figures in column 3–5 and 7–8 are reconstructed boundaries from Theorem 1.

Table 1. Reconstruction results for both the greedy algorithm with different starting points and the ILP algorithm.

Original	Noisy	Greedy local algorithm				Greedy + ILP	ILP algorithm
		First s. p.	Second s. p.	Third s. p.	Superposition	Superpos.+ILP	ILP solution

7   Discussion

The ILP algorithm reconstruction to the far right in Table 1 works well, although there are more grey pixels in the reconstructed image than in the original one. This may partly be due to boundary effects, since configurations near the boundary have fewer neighbour configurations they need to match.

The main problem with the ILP algorithm is that the NP-hardness makes it very time-consuming. Several tests indicate that for images only slightly bigger than those in this paper, the ILP algorithm is so slow that it is of no practical use.

The running time for the greedy algorithm is $O(n^4 \log_2 n)$ for an $n \times n$ image, but images with many white pixels are processed faster than images with many black pixels. In practical cases, the greedy algorithm is much faster and therefore easier to work with. However, results from the greedy algorithm are generally not as nice as results from the ILP algorithm. The output quality depends greatly on the starting point and the algorithm is not good at finding large black areas or the right number of components. This is due to the construction of the algorithm, which may still be improved. The algorithm can also not reconstruct loops, since this requires addition of a configuration not in the list C.

All of these flaws are present in the outputs of the greedy algorithm in Table 1. However, the superposition is a good approximation of the object, and when the ILP and greedy algorithms are used together, we get a good suggestion for the reconstruction as seen in column 7 of the table. Empirical results seem to indicate that the running time of the ILP algorithm on the superposition of the outputs from the greedy algorithm is a bit shorter than the running time of the ILP algorithm used directly on the noisy image, but note that the ILP algorithm used with the greedy algorithm is still rather slow since the ILP is still solving an NP-hard problem.

To sum up, the ILP algorithm gives the best output, but its running time must be brought down if it is to be of practical relevance. The greedy algorithm is faster, but the quality of the output is less reliable. We are still working on improving running time and output for both algorithms.

References

1. Caselles, V., Kimmel, R., Sapiro, G.: Geodesic active contours. Int. J. Comput. Vis. **22**(1), 61–79 (1997)
2. Chan, T., Vese, L.: Active contours without edges. IEEE TIP **10**(2), 266–277 (2001)
3. Cootes, T.F., Taylor, C.J.: Active shape models - "smart snakes". In: Hogg, D., Boyle, R. (eds.) BMVC92. Springer, London (1992). https://doi.org/10.1007/978-1-4471-3201-1_28
4. Duarte, P., Torres, M.J.: Smoothness of boundaries of regular sets. J. Math. Imaging Vis. **48**(1), 106–113 (2014)
5. Geman, S., Geman, D.: Stochastic relaxation, Gibbs distributions, and the Bayesian restoration of images. TPAMI **6**(6), 721–741 (1984)
6. Kass, M., Witkin, A., Terzopoulos, D.: Snakes: active contour models. IJCV **1**(4), 321–331 (1988)

7. Latecki, L., Conrad, C., Gross, A.: Preserving topology by a digitization process. J. Math. Imaging Vis. **8**, 131–159 (1998)
8. Litjens, G., et al.: A survey on deep learning in medical image analysis. MedIA **42**, 60–88 (2017)
9. Mumford, D., Shah, J.: Optimal approximations by piecewise smooth functions and associated variational problems. Comm. Pure Appl. Math. **42**, 577–684 (1989)
10. Pavlidis, T.: Algorithms for Graphics and Image Processing. Digital System Design Series. Springer, Heidelberg (1982). https://doi.org/10.1007/978-3-642-93208-3
11. Ronse, C., Tajine, M.: Discretization in Hausdorff space. J. Math. Imaging Vis. **12**(3), 219–242 (2000)
12. SDSS, S.D.S.S. https://www.sdss.org
13. Serra, J.: Image Analysis and Mathematical Morphology. Academic Press Inc., Orlando (1983)
14. Stelldinger, P., Köthe, U.: Shape preservation during digitization: tight bounds based on the morphing distance. In: Michaelis, B., Krell, G. (eds.) DAGM 2003. LNCS, vol. 2781, pp. 108–115. Springer, Heidelberg (2003). https://doi.org/10.1007/978-3-540-45243-0_15
15. Stelldinger, P., Köthe, U.: Towards a general sampling theory for shape preservation. Image Vision Comput. **23**(2), 237–248 (2005)
16. Svane, H., du Plessis, A.: Reconstruction of r-regular objects from trinary images, to be published in Ph.D thesis of H. Svane (2019). Until then available on Arxiv
17. V12.8.0, I.I.C.O.S. www.cplex.com
18. Yezzi, A., Kichenassamy, S., Kumar, A., Olver, P., Tannenbaum, A.: A geometric snake model for segmentation of medical imagery. IEEE TMI **16**(2), 199–209 (1997)

Network Embedding by Walking on the Line Graph

Miguel Angel Lozano[1](ID), Manuel Curado[1](ID), Francisco Escolano[1(✉)](ID),
and Edwin R. Hancock[2](ID)

[1] Department of Computer Science and AI, University of Alicante, Alicante, Spain
malozano@ua.es, {mcurado,sco}@dccia.ua.es
[2] Department of Computer Science, University of York, York, UK
edwin.hancock@york.ac.uk

Abstract. In this paper, we propose to embed edges instead of nodes using state-of-the-art neural/factorization methods (DeepWalk, node2vec). These methods produce latent representations based on co-ocurrence statistics by simulating fixed-length random walks and then taking bags-of-vectors as the input to the Skip Gram Learning with Negative Sampling (SGNS). We commence by expressing commute times embedding as matrix factorization, and thus relating this embedding to those of DeepWalk and node2vec. Recent results showing formal links between all these methods via the spectrum of graph Laplacian, are then extended to understand the results obtained by SGNS when we embed edges instead of nodes. Since embedding edges is equivalent to embedding nodes in the line graph, we proceed to combine both existing formal characterizations of the line graphs and empirical evidence in order to explain why this embedding dramatically outperforms its nodal counterpart in multi-label classification tasks.

Keywords: Network embedding · SGNS · Line graph · Spectral theory

1 Introduction

The recent success of neural graph embeddings such as LINE [18], DeepWalk [14] and node2vec [7] has opened a new path for analyzing networks. Despite these embeddings outperform spectral ones in tasks such as link prediction and multi-label node classification, Spectral Graph Theory [3] is still key tool for understanding and characterizing neural embeddings [16].

In this paper, we contribute with empirical evidence showing that neural embeddings (Sect. 2) can boost their performance in multi-label classification by embedding edges instead of nodes. We conjecture that this fact is due to the spectral properties of *line graphs*, whose nodes are the edges of the original graphs

M. A. Lozano, M. Curado and F. Escolano are funded by the project TIN2015-69077-P of the Spanish Government.

(Sect. 3). However, since general line graphs have not been fully characterized yet, we can only correlate our empirical findings (Sect. 4) with some of the well known properties of line graphs and the spectral characterization of neural embeddings.

2   Classic vs Neural Embeddings

2.1   Classic Embeddings

Let $G = (V, E, \mathbf{A})$ be a graph/network with $n = |V|$ nodes, $m = |E|$ edges, where $E \subseteq V \times V$, and adjacency matrix $\mathbf{A}$. Then, *node embedding* consists of finding a mapping $f : V \to \mathbb{R}^d$ (with $d \ll n$) so that the resulting d-dimensional vectors capture the structural properties of each vertex. As a result, we have $||f(i) - f(j)||^2 \to 0$ if nodes i and j are structurally similar within the graph G. Traditionally, nodal structural similarity was associated with the reachability of node j from node i (and vice versa) through random walks [10]. This characterization leaded to define both *hitting times* H_{ij} (expected steps taken by a random walk to reach j from i) and *commute times* $CT_{ij} = H_{ij} + H_{ji}$ (which also includes the expected steps needed to return to i from j). Since random walks are encoded by transition matrices of the form $\mathbf{P} = \mathbf{D}^{-1}\mathbf{A}$, where $\mathbf{D} = diag(d_1, \ldots, d_n)$ is the diagonal matrix with the degrees of the nodes, the spectral analysis of $\mathbf{P}$ is a natural way of understanding both hitting and commute times. More precisely, let $\lambda_1 = 1 \geq \lambda_2 \geq \ldots \geq \lambda_n \geq -1$ be the spectrum of the transition matrix. It is well known that hitting times and commute times are highly conditioned by the *spectral gap* $\lambda = 1 - \max\{\lambda_2, |\lambda_n|\}$. When several communities are encoded by a connected graph G, then H_{ij} and CT_{ij} are only meaningful when $\lambda \to 0$ (small bottlenecks between communities); otherwise, these quantities rely on the local densities (degrees) of the nodes i and j, and one cannot discriminate whether two nodes belong to the same community or not [11]. Consequently, the applicability of node embeddings based on commute times to clustering is quite limited (see representative examples of image segmentation and tracking in [15]). In this regard, recent research is focused on simultaneously minimizing the spectral gap and shrinking (whenever possible) inter-community commute distances via graph densification [4,5] before embedding the nodes.

Therefore, once G is processed (or rewired) commute times embedding leads to learn two matrices $\mathbf{X}, \mathbf{Y} \in \mathbb{R}^{n \times d}$, whose rows are denoted by $\mathbf{x}_i$ and $\mathbf{y}_i$ respectively and $\mathbf{x}_i$ is the embedding of the node i. Following [15], the commute times embedding matrix $\mathbf{X}$ results from factorizing

$$vol(G)\mathcal{G} = \mathbf{X}\mathbf{Y}^T, \tag{1}$$

where $vol(G) = \sum_{i=1}^n d_i$ is the volume of the graph and $\mathcal{G}$ is its Green's function, i.e. the pseudo-inverse of the normalized graph Laplacian $\mathcal{L} = \mathbf{I} - \mathbf{D}^{-1/2}\mathbf{A}\mathbf{D}^{-1/2}$, whose spectrum is $1 - \lambda_1 = 0, 1 - \lambda_2, \ldots, 1 - \lambda_n \leq 2$, i.e. if λ_i is an eigenvalue of $\mathbf{P}$ then $1 - \lambda_i$ is an eigenvalue of $\mathcal{L}$.

2.2   Neural Embeddigs

Neural embeddings such as LINE [18], DeepWalk [14] and node2vec [7], exploit random walks in a different way. Namely, they simulate a fixed number N of random walks with fixed length L emanating from the nodes of G and then capture co-ocurrence statistics of pairs of nodes. The first node of the i-th path w_i, assimilated to a word in a textual corpus (skip-gram model), is sampled from a prior distribution $P(w_i)$. Then, the context of w_i is given by the nodes/words surrounding it in a T-sized window $w_{i-T}, \ldots, w_{i-1}, w_{i+1}, \ldots, w_{i+T}$, according to the transition matrix $\mathbf{P}$. Then, the node-context pairs (w, c) are given by (w_{i-r}, w_i) and (w_i, w_{i+r}) for $r = 1, \ldots, T$. All these pairs are added to the multiset $\mathcal{D}$ used for learning with *negative sampling*. Negative sampling implies not only to consider likely node-context pairs (w, c) but also b unlikely ones (w, c'): the negative samples c', are nodes that can be drawn from the steady-state probability distribution of the random walk, i.e. $P_N(i) = d_i/vol(G)$. This process is called Skip Gram Learning with Negative Sampling (SGNS) and leads to the following factorization [9]:

$$\mathbf{M} = \mathbf{XY}^T, \text{ with } \mathbf{M}_{ij} = \log\left(\frac{\#(w_i, c_j)|\mathcal{D}|}{\#(w_i)\#(c_j)}\right) - \log b, \qquad (2)$$

where: $\#(w_i, c_i)$ is the number of times the corresponding node-context pair is observed, $\#(w_i)$ is the number of times the node i is observed and similarly for node $\#(c_j)$; finally $\log(.)$ is the element-wise logarithm and b is the number of negative samples.

2.3   LINE and DeepWalk vs node2vec Factorizations

These strategies differ in the way they sample (and thus vectorize) the graph for SGNS. LINE and DeepWalk rely on first-order random walks whereas node2vec is driven by second-order random walks.

LINE and DeepWalk. LINE's factorization is a direct result from the cost function associated with SGNS. In particular, the latent representations of both the word/node $\mathbf{x}_i$ and the context $\mathbf{y}_j$ are assumed to be correlated with the existence on an edge between nodes i and j, i.e. $\mathbf{A}_{ij} \log g(\mathbf{x}_i^T \mathbf{y}_j)$ is maximized, where $g(.)$ is the sigmoid function. Following [16], this leads to

$$\mathbf{x}_i^T \mathbf{y}_j = \log\left(\frac{vol(G)\mathbf{A}_{ij}}{d_i d_j}\right) - \log b \Rightarrow \log\left(vol(G)\mathbf{D}^{-1}\mathbf{A}\mathbf{D}^{-1}\right) - \log b = \mathbf{XY}^T. \tag{3}$$

DeepWalk, on the other hand, leads to a more complex factorization. Assuming that the first node of each random walk is drawn from the steady state distribution, we have that, when $L \to \infty$,

$$\frac{\#(w_i, c_j)|\mathcal{D}|}{\#(w_i)\#(c_j)} \xrightarrow{p} \frac{vol(G)}{2T}\left(\frac{1}{d_j}\sum_{r=1}^{T}\mathbf{P}_{ij}^r + \frac{1}{d_i}\sum_{r=1}^{T}\mathbf{P}_{ji}^r\right) \tag{4}$$

where $\xrightarrow{p}$ denotes *convergence in probability*. This yields

$$\log\left(\frac{vol(G)}{T}\left(\sum_{r=1}^{T}\mathbf{P}^r\right)\mathbf{D}^{-1}\right) - \log b = \mathbf{X}\mathbf{Y}^T, \tag{5}$$

which is equivalent to LINE for $T = 1$.

node2vec. The underlying idea of this embedding is to add more flexibility to the random walk. This is done by defining two parameters p and q, that control, respectively the likelihood of immediately revisit a node in the walk and making the walk very local. To that end, node2vec needs to evaluate the probability of the next nodes given the preceding one in the walk, i.e. we have a 2nd-order random walk. This walk is characterized by the hypermatrix $\mathbf{P}$, where $\mathbf{P}_{i(jk)}$ denotes the probability of reaching j from j given that the node preceeding j is k. Thus, the 2nd order random walk can be reduced to a 1st order one on the edges of the graph [1] as it is done in the implementation of node2vec. The stationary distribution $\mathbf{X}_{ik}$ for this type of random walks satisfies $\sum_k \mathbf{P}_{i(jk)}\mathbf{X}_{ik} = \mathbf{X}_{ij}$. Qiu et al. [16] have found that

$$\frac{\#(w_i, c_j)|\mathcal{D}|}{\#(w_i)\#(c_j)} \xrightarrow{p} \frac{\frac{1}{2T}\sum_{r=1}^{T}\left(\sum_k \mathbf{X}_{ik}\mathbf{P}_{j(ik)}^r + \sum_k \mathbf{X}_{jk}\mathbf{P}_{i(jk)}^r\right)}{\left(\sum_k \mathbf{X}_{ik}\right)\left(\sum_k \mathbf{X}_{jk}\right)} \tag{6}$$

and, despite the matricial expression for the factorization is more elusive, the final factorization differs significantly from those of DeepWalk and LINE.

3   Node vs Edges Embedding

3.1   The Line Graph

In this paper, we are mainly concerned with the impact of embedding the edges of G instead of its nodes. This means that a word w_i in the previous expressions is not yet associated with a node of G but with a node of its *line graph* L_G. The nodes of L_G are the edges of G and there is an edge in the line graph if two edges in G share a node. More formally, given the $n \times m$ incidence matrix $\mathbf{B}$ where $\mathbf{B}_{i\alpha}$ is 1 if the link α is related to node i and 0 otherwise, we have that the $m \times m$ adjacency matrix $\mathbf{C} = \mathbf{B}^T\mathbf{B} - \mathbf{I}$ has elements $C_{\alpha\beta} = \sum_{i=1}^{n}\mathbf{B}_{i\alpha}\mathbf{B}_{i\beta}(1 - \delta_{\alpha\beta})$.

3.2   Spectral Analysis

Some interesting properties of line graphs vs G:

- *Boosted edge density.* A single node i in G leads to a clique of $d_i(d_i - 1)/2$ edges in L_G (see Fig. 1). Despite this gives a high prominence to notable nodes of G it flexibilizes community detection [6]. In addition the steady state distribution of a random walk in L_G is $P_N(\alpha_{(i,j)}) = d_\alpha/vol(L_G)$ where $d_\alpha = d_i + d_j - 2$ and $vol(L_G) = \sum_{\alpha,\beta} C_{\alpha\beta} = \sum_{i=1}^{n} d_i(d_i - 1)$.

- *Redundant spectrum for $m > n$.* Let $\lambda_1(L_G) \geq \lambda_2(L_G) \geq \ldots \geq \lambda_m(L_G)$ be the spectrum of $\mathbf{C}$. Then, for $m > n$, $\lambda_{n+1} = \ldots = \lambda_m = -2$. As a result, $\lambda_i(\mathbf{L}(L_G)) \geq 4$, for the largest $m - n$ eigenvalues of $\mathbf{L}(L_G)$, the unnormalized Laplacian matrix of L_G [19]. This increases significantly the medium-large eigenvalues of $\mathcal{L}(L_G)$ with respect to $\mathcal{L}(G)$ (see Fig. 2).
- *Majorization of the spectrum of G.* This is really a conjecture derived from the bound $\lambda_2(L_G) \leq \frac{m}{2} - 1$ in comparison to that for G: $\lambda_2(G) \leq \frac{n}{2} - 1$. Empirical data shows that the lowest part of the spectrum of $\mathcal{L} - \mathbf{I}$ in the line graph majorizes that of G (blue lines in Fig. 2). Since the spectrum driving DeepWalk is (approximately) of the form $\frac{1}{T} \sum_{r=1}^{T} \lambda_i^r$ this leads (in general) to small spectral gaps for the line graphs, and thus slower mixing times of the random walks (more randomness). Green lines show the real spectra driving random walks in DeepWalk. In all cases, $T = 10$.

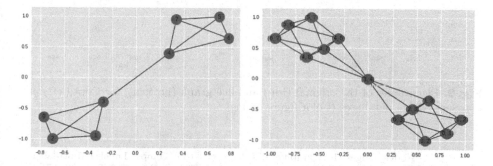

Fig. 1. Barbell graph linking two cliques (left) and its line graph (right)

4   Experiments and Discussion

4.1   Datasets (Networks)

- **CiteSeer for Document Classification** [17]. Citation network containing 3312 scientific publications with 4676 links between them. Each publication is classified in one of 6 categories.
- **Cora** [17]. Citation network containing 2708 scientific publications with 5278 links between them. Each publication is classified in one of 7 categories.
- **Wiki**[1]. Contains a network of 2405 web pages with 17981 links between them. Each page is classified in one of 19 categories.
- **Facebook** social circles [13].

---

[1] https://github.com/thunlp/MMDW/.

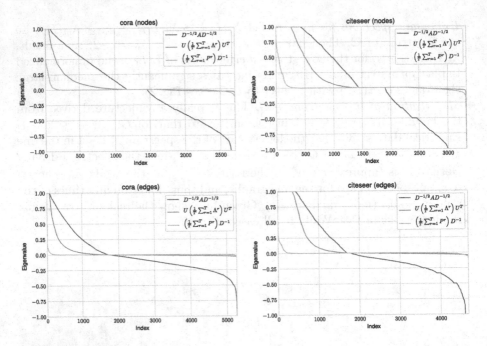

Fig. 2. Eigenvalues of the original (top) and line graph (bottom), for Cora (left) and CiteSeer (right) databases (Color figure online)

	Nodes	Edges	Line graph edges	Gap	Con. comps	Labels	Multi-label
wiki	2405	12761	355644	0.000000	45	19	No
cora	2708	5278	52301	0.000000	78	7	No
citeseer	3327	4676	27174	0.000000	438	6	No
ppi	3890	38739	3018220	0.000000	35	50	Yes
pos	4777	92517	49568882	0.576132	1	40	Yes
facebook	4039	88234	9314849	0.000837	1	10	Yes

- **Wikipedia Part-of-Speech (POS)** [12]. Co-ocurrence of words appearing in the first million of the bytes of the dumping of Wikipedia. The categories correspond to the labels of Part-of-Speech (POS) inferred by the Stanford POS-Tagger. Contains 4777 nodes, and 92517 undirected links. Each node may have several labels. We have 40 labels (categories).
- **Protein-Protein Interactions (PPI)**[2] [2]. We use a subgraph of the PPIs associated with the Homo Sapiens. The network has 3890 nodes and 76584 links. Each node may have several labels corresponding to the 50 possible categories.

---

[2] https://downloads.thebiogrid.org/BioGRID.

Facebook, PPI and POS have been retrieved from SNAP[3] [8]. CiteSeer and Cora have been retrieved from LINQS[4].

All the networks are considered as undirected graphs. Originally single-labelled networks are transformed into multi-label networks when their line-graph is computed (or sampled, for the sake of efficiency). Nodes with more than one label are the *border nodes* between two categories (inter-class), and the nodes that hold one label are intra-class nodes.

	Inter-class nodes	Intra-class nodes
wiki	4526 (35%)	8235 (65%)
cora	1003 (19%)	4275 (81%)
citeseer	1190 (25%)	3486 (75%)

We have used the implementations of node2vec and DeepWalk included in the framework. OpenNE[5]. The default values for p and q in node2vec are $p = q = 1$. After optimizing p and q in the range $\{0.25, 0.5, 1, 2, 4\}$ the maximum improvement of node2vec wrt DeepWalk in the classification score is 0.014 (micro and macro). Regarding spectral embeddings, CTE and LLE have been only tested in networks with a single connected component (pos and facebook). In particular, commute times (CTE) have a poor performance in multi-label classification because their factorization relies on the Green's function and this means that only the inverse of each eigenvalue is considered. However, DeepWalk is controlled by a polynomial associated with each eigenvalue.

		node2vec		DeepWalk		cte	lle
		Nodes	Edges	Nodes	Edges		
Micro-F1	citeseer	0.591071	0.768626	0.595833	**0.780930**	–	–
	cora	0.808715	0.903557	0.817578	**0.920538**	–	–
	wiki	0.663342	0.840709	0.692436	**0.859129**	–	–
	pos	0.447845	**0.697572**	0.471620	0.696640	0.375037	0.393165
	ppi	0.197435	0.590731	0.205786	**0.609937**	–	–
	facebook	0.911516	**0.999900**	0.911516	**0.999900**	0.239217	0.231214

---

[3] https://snap.stanford.edu/node2vec/.
[4] https://linqs.soe.ucsc.edu/data.
[5] https://github.com/thunlp/OpenNE.

| | | node2vec | | DeepWalk | | cte | lle |
		Nodes	Edges	Nodes	Edges		
Macro-F1	citeseer	0.544490	0.730930	0.545774	**0.745995**	–	–
	cora	0.798782	0.898863	0.804928	**0.917838**	–	–
	wiki	0.528603	0.764205	0.597948	**0.787771**	–	–
	pos	0.084183	0.773035	0.094148	**0.774001**	0.041637	0.033617
	ppi	0.168237	0.566912	0.178405	**0.587934**	–	–
	facebook	0.821899	0.999377	0.822243	**0.999407**	0.108742	0.045155

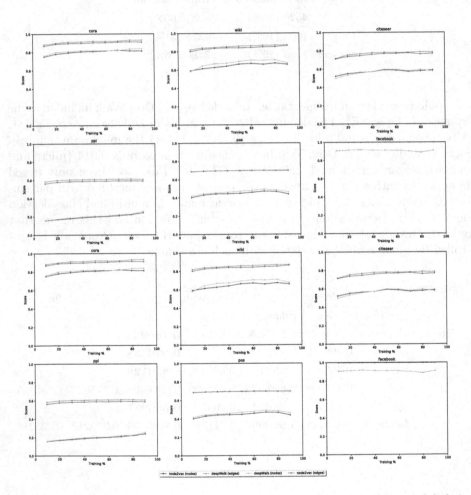

Fig. 3. Evolution of the performance as a function of the fraction of known labels in the training set. Micro-F1 (top) and Macro-F1 (bottom)

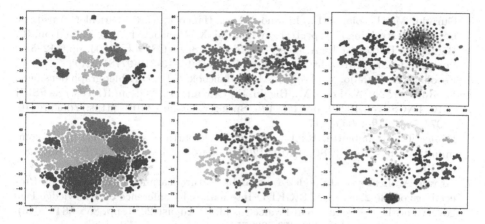

Fig. 4. t-SNE embeddings. Original graph (top) and line-graph (bottom), for Facebook (left), Cora (center) and CitcSeer (right) databases.

In Fig. 3, we show the performance in multi-label classification (according to the percentage of nodes with known labels). The line-graph versions of node2vec and DeepWalk clearly outperform their nodal counterparts. The similarity in terms of performance of node2vec and DeepWalk is due to the fact that the 2nd order random walk of node2vec is not applied at the level of edges (it is unfeasible for large networks). Finally, in Fig. 4 we show the t-SNE embeddings. Edge embeddings clearly produce denser communities.

5   Conclusions

In this paper, we have contributed with empirical evidence showing that embedding edges clearly outperforms node-based embeddings in neural SGNS strategies. We conjecture that this is due to the slower mixing times of random walks in line graphs. Future work includes a detailed check of this conjecture as well as more efficient (in time and space) strategies for designing walkers on the line graphs.

References

1. Benson, A.R., Gleich, D.F., Lim, L.: The spacey random walk: a stochastic process for higher-order data. SIAM Rev. **59**(2), 321–345 (2017). https://doi.org/10.1137/16M1074023
2. Breitkreutz, B., et al.: The biogrid interaction database: 2008 update. Nucleic Acids Res. **36**, 637–640 (2008). https://doi.org/10.1093/nar/gkm1001
3. Chung, F.R.K.: Spectral graph theory. In: Conference Board of the Mathematical Sciences (CBMS), number 92. American Mathematical Society (1997)

4. Curado, M., Escolano, F., Lozano, M.A., Hancock, E.R.: Dirichlet densifiers: beyond constraining the spectral gap. In: Bai, X., Hancock, E., Ho, T., Wilson, R., Biggio, B., Robles-Kelly, A. (eds.) S+SSPR 2018. LNCS, vol. 11004, pp. 512–521. Springer, Cham (2018). https://doi.org/10.1007/978-3-319-97785-0_49

5. Escolano, F., Curado, M., Lozano, M.A., Hancook, E.R.: Dirichlet graph densifiers. In: Robles-Kelly, A., Loog, M., Biggio, B., Escolano, F., Wilson, R. (eds.) S+SSPR 2016. LNCS, vol. 10029, pp. 185–195. Springer, Cham (2016). https://doi.org/10. 1007/978-3-319-49055-7_17

6. Evans, T.S., Lambiotte, R.: Line graphs, link partitions, and overlapping communities. Phys. Rev. E **80**, 016105 (2009). https://doi.org/10.1103/PhysRevE.80. 016105

7. Grover, A., Leskovec, J.: node2vec: scalable feature learning for networks. In: Proceedings of the 22nd ACM SIGKDD International Conference on Knowledge Discovery and Data Mining, San Francisco, 13–17 August 2016, pp. 855–864 (2016). https://doi.org/10.1145/2939672.2939754

8. Leskovec, J., Krevl, A.: SNAP datasets: Stanford large network dataset collection, June 2014. http://snap.stanford.edu/data

9. Levy, O., Goldberg, Y.: Neural word embedding as implicit matrix factorization. In: Advances in Neural Information Processing Systems 27: Annual Conference on Neural Information Processing Systems, 8–13 December 2014, Montreal, pp. 2177–2185 (2014). http://papers.nips.cc/paper/5477-neural-word-embedding-as-implicit-matrix-factorization

10. Lovász, L.: Random walks on graphs: a survey. In: Miklós, D., Sós, V.T., Szőnyi, T. (eds.) Combinatorics, Paul Erdős is Eighty, vol. 2, pp. 353–398. János Bolyai Mathematical Society, Budapest (1996)

11. von Luxburg, U., Radl, A., Hein, M.: Hitting and commute times in large random neighborhood graphs. J. Mach. Learn. Res. **15**(1), 1751–1798 (2014). http://dl.acm.org/citation.cfm?id=2638591

12. Mahoney, M.: Large text compression benchmark (2011). http://www. mattmahoney.net/dc/textdata

13. McAuley, J.J., Leskovec, J.: Learning to discover social circles in ego networks. In: Advances in Neural Information Processing Systems 25: 26th Annual Conference on Neural Information Processing Systems, Proceedings of a Meeting Held 3–6 December 2012, Lake Tahoe, pp. 548–556 (2012). http://papers.nips.cc/paper/ 4532-learning-to-discover-social-circles-in-ego-networks

14. Perozzi, B., Al-Rfou, R., Skiena, S.: DeepWalk: online learning of social representations. In: The 20th ACM SIGKDD International Conference on Knowledge Discovery and Data Mining, KDD 2014, New York, 24–27 August 2014, pp. 701–710 (2014). https://doi.org/10.1145/2623330.2623732

15. Qiu, H., Hancock, E.R.: Clustering and embedding using commute times. IEEE Trans. Pattern Anal. Mach. Intell. **29**(11), 1873–1890 (2007). https://doi.org/10. 1109/TPAMI.2007.1103

16. Qiu, J., Dong, Y., Ma, H., Li, J., Wang, K., Tang, J.: Network embedding as matrix factorization: unifying DeepWalk, LINE, PTE, and node2vec. In: Proceedings of the Eleventh ACM International Conference on Web Search and Data Mining, WSDM 2018, pp. 459–467. ACM, New York (2018). https://doi.org/10. 1145/3159652.3159706

17. Sen, P., Namata, G., Bilgic, M., Getoor, L., Gallagher, B., Eliassi-Rad, T.: Collective classification in network data. AI Mag. **29**(3), 93–106 (2008). http://www.aaai.org/ojs/index.php/aimagazine/article/view/2157
18. Tang, J., Qu, M., Wang, M., Zhang, M., Yan, J., Mei, Q.: LINE: large-scale information network embedding. In: Proceedings of the 24th International Conference on World Wide Web, WWW 2015, Florence, 18–22 May 2015, pp. 1067–1077 (2015). https://doi.org/10.1145/2736277.2741093
19. Yan, C.: Properties of spectra of graphs and line graphs. Appl. Math. J. Chin. Univ. **17**(3), 371–376 (2002). https://doi.org/10.1007/s11766-002-0017-7

Discriminant Manifold Learning with Graph Convolution Based Regression for Image Classification

Ruifeng Zhu[1,2], Fadi Dornaika[2,3(✉)], and Yassine Ruichek[1]

[1] Laboratory of Electronics, Information and Image (LE2I), CNRS,
University of Bourgogne Franche-Comte (UBFC), Belfort, France
[2] University of the Basque Country UPV/EHU, San Sebastian, Spain
`fadi.dornaika@ehu.es`
[3] IKERBASQUE, Basque Foundation for Science, Bilbao, Spain

Abstract. Many learning problems can be cast into learning from data-driven graphs. This paper introduces a framework for supervised and semi-supervised learning by estimating a non-linear embedding that incorporates Spectral Graph Convolutions structure. The proposed algorithm exploits data-driven graphs in two ways. First, it integrates data smoothness over graphs. Second, the regression is solved by the joint use of the data and their graph in the sense that the regressor sees convolved data samples. The resulting framework can solve the problem of over-fitting on local neighborhood structures for image data having varied natures like outdoor scenes, faces and man-made objects. Our proposed approach not only provides a new perspective to non-linear embedding research but also induces the standpoint on Spectral Graph Convolutions methods. In order to evaluate the performance of the proposed method, a series of experiments are conducted on four image datasets in order to compare the proposed method with some state-of-art algorithms. This evaluation demonstrates the effectiveness of the proposed embedding method.

Keywords: Graph-based embedding · Supervised learning · Semi-supervised learning · Spectral graph convolutions · Discriminative embedding

1 Introduction

Graph-based supervised and semi-supervised learning can be used by many real world applications. Graph-based methods exploits data structure in discovering the sought models. Besides, these approaches can encode the relationship between the nodes of labeled or unlabeled samples data. They can be used for reducing the dimensionality of data.

In graph-based learning field, several effective algorithms were developed in the past: Locally linear embedding (LLE) [16], Laplacian Eigenmap (LE) [1] and

© Springer Nature Switzerland AG 2019
D. Conte et al. (Eds.): GbRPR 2019, LNCS 11510, pp. 226–236, 2019.
https://doi.org/10.1007/978-3-030-20081-7_22

ISOMAP [17] are classical graph-based non-linear embedding algorithms. Flexible Semi-Supervised Embedding algorithm (FSSE) was proposed in [4]. This method simultaneously estimates a non-linear embedding and its linear regressor. Manifold Regularized Deep Learning Architecture Algorithm (MRDL) [19] proposed a deep architecture to learn the high-level features for scene recognition in an unsupervised fashion. In [7], the authors proposed a joint embedding learning and sparse regression (JELSR) for unsupervised feature selection. Semi-Supervised Discriminant Embedding (SDE) [18] is the semi-supervised extension of Local Discriminant Embedding (LDE) [3]. It is a linear projection method that is based on manifold smoothness and a regularizer that controls the complexity of learning. Semi-Supervised Discriminant Analysis (SDA) [2] extends the classic Linear Discriminant Analysis (LDA) [12] by adding a geometrically-based regularization term in the objective function of LDA.

This paper introduces a framework for supervised and semi-supervised learning by estimating a flexible non-linear embedding that incorporates Spectral Graph Convolutions structure. The proposed algorithm exploits data-driven graphs in two ways. First, it integrates data smoothness over graphs. Second, the regression is solved by the joint use of the data and their graph in the sense that the regressor sees convolved data samples.

The main contributions of the paper are as follows:

- We propose an unified framework for the non-linear embedding. This framework integrates manifold smoothness, large margin concept, and sparse regression for image classification. Besides the jointly estimated regressor function utilizes the Graph Spectral Convolutions, where the graph is sparse data-driven.
- We provide a closed-form solution to the optimization problem.
- The proposed framework can be used by both settings: supervised and semi-supervised.
- Extensive experiments demonstrate that the proposed non-linear embedding method can be superior to many state of the art graph-based methods.

The rest of the paper is structured as follows. Section 2 introduces the notations and definitions used in this paper, and briefly reviews some related works. Section 3 presents the proposed framework for non-linear embedding. Section 4 reports and analyzes some experimental results obtained with four public image datasets. Finally, Sect. 5 concludes the paper. In this paper, capital bold letters denote matrices and small bold letters denote vectors.

2   Related Work

2.1   Notation and Preliminaries

In this section, we introduce the notations adopted in the paper. We define the data matrix by $\mathbf{X} = [\mathbf{x}_1, \mathbf{x}_2, \cdots, \mathbf{x}_l, \mathbf{x}_{l+1}, \cdots, \mathbf{x}_{l+u}] \in \mathbb{R}^{d \times (l+u)}$, where $\mathbf{x}_i \big|_{i=1}^{l}$ and $\mathbf{x}_i \big|_{i=l+1}^{l+u}$ are the labeled training samples and unlabeled test samples, respectively, l and u being the total numbers of labeled train samples and unlabeled test samples data, respectively, and d being the sample dimension. Let $N = l + u$ be the total number of samples data and n_c be the total number of labeled samples in the cth class. The labeled train samples are denoted by the matrix $\mathbf{X}_l = [\mathbf{x}_1, \mathbf{x}_2, \cdots, \mathbf{x}_l] \in \mathbb{R}^{d \times l}$. The label of each sample $\mathbf{x}_i$ is denoted by $y_i \in (1, 2, ..., C)$, where C is the total number of classes. The unlabeled (test) samples are denoted by the matrix $\mathbf{X}_u = [\mathbf{x}_{l+1}, \mathbf{x}_{l+2}, \cdots, \mathbf{x}_{l+u}] \in \mathbb{R}^{d \times u}$. Let $\mathbf{S} \in \mathbb{R}^{(l+u) \times (l+u)}$ be the graph similarity matrix associated with the data matrix $\mathbf{X}$ where $S(i, j)$ represents the similarity between $\mathbf{x}_i$ and $\mathbf{x}_j$, i.e., $S(i, j) = sim(\mathbf{x}_i, \mathbf{x}_j)$. The function $sim(.,.)$ can be any symmetric function that measures the similarity between two samples. This can be given by the Cosine or the Gaussian Kernel.

The Laplacian of the similarity matrix $\mathbf{S}$ is denoted by $\mathbf{L}$ and is given by $\mathbf{L} = \mathbf{D} - \mathbf{S}$ where $\mathbf{D}$ is a diagonal matrix whose elements are the row or column (since the matrix is symmetric) sums of $\mathbf{S}$ matrix. The normalized Laplacian matrix $\mathbf{L}$ is defined by $\mathbf{L} = \mathbf{I} - \mathbf{D}^{-1/2} \mathbf{S} \mathbf{D}^{1/2}$ where $\mathbf{I}$ denotes the identity matrix. The normalized Laplacian matrix $\mathbf{L}$ could be decomposed by its eigenvectors matrix $\mathbf{U}$, $\mathbf{L} = \mathbf{I} - \mathbf{D}^{-\frac{1}{2}} \mathbf{S} \mathbf{D}^{-\frac{1}{2}} = \mathbf{U} \mathbf{\Lambda} \mathbf{U}^T$, where $\mathbf{\Lambda}$ means a diagonal matrix of eigenvalues of $\mathbf{L}$.

2.2   Manifold Regularized Deep Learning Algorithm (MRDL)

The Manifold Regularized Deep Learning Algorithm (MRDL) in [19] estimates a Non-linear Sparsity Preserving Projection [9]. It adopts a cascade of layers. In each layer, a sparse graph is computed and then the non-linear projections are estimated using sparsity preserving criterion.

The objective function allowing the estimation of the similarity (graph) matrix $\mathbf{S}$ is:

$$\min_{\mathbf{S} \geq 0} \|\mathbf{X} - \mathbf{X} \mathbf{S}\|_2^2 + \alpha \|\mathbf{S}\|^{\frac{1}{2}} + \beta \|\mathbf{S}\|_2^2 \tag{1}$$

where α and β are two positive regularization parameters.

Once the graph matrix $\mathbf{S}$ is computed, the non-linear projections $\mathbf{Y} = (\mathbf{y}_1, \mathbf{y}_2, \cdots \mathbf{y}_n)$ can be estimated by solving the following optimization problem:

$$\min_{\mathbf{Y}} \sum_i \left\| \mathbf{y}_i - \frac{1}{2} \sum_j (\mathbf{D}^{-1}(\mathbf{S}^T + \mathbf{S}))_{ij} \mathbf{y}_j \right\|^2$$
$$= \min_{\mathbf{Y}} tr(\mathbf{Y} \mathbf{D}^{\frac{1}{2}} (\mathbf{I} - \tilde{\mathbf{S}})^2 \mathbf{D}^{\frac{1}{2}} \mathbf{Y}^T)$$
$$\Rightarrow \min_{\mathbf{G}} tr(\mathbf{G}^T (\mathbf{I} - \tilde{\mathbf{S}})^2 \mathbf{G}) \tag{2}$$

where $\tilde{\mathbf{S}} = \frac{1}{2}(\mathbf{S} + \mathbf{S}^T) \mathbf{D}^{-1}$, $\mathbf{G} = \mathbf{D}^{\frac{1}{2}} \mathbf{Y}^T$. The optimal solution for $\mathbf{G}$ (or equivalently $\mathbf{Y}$) is given by the smallest eigenvectors of $(\mathbf{I} - \tilde{\mathbf{S}})$.

2.3   Graph Convolutional Networks (GCN)

The Graph Convolutional Networks(GCN) algorithm in [8] presents an approach for semi-supervised learning on graph-structured data that is based on an efficient variant of convolutional neural networks which operate directly on graphs.

For semi-supervised label propagation, a two-layer GCN model has the following form:

$$\mathbf{E} = f(\mathbf{X}, \mathbf{A}) = \text{softmax}(\hat{\mathbf{A}}\text{ReLU}(\hat{\mathbf{A}}\dot{\mathbf{X}}\mathbf{W}^{(0)})\mathbf{W}^{(1)}) \tag{3}$$

where $\dot{\mathbf{X}} \in \mathbb{R}^{\mathbf{N} \times \mathbf{d}}$ denotes the input data with N samples and d dimensions, $\hat{\mathbf{A}}$ is a renormalized graph matrix, $\mathbf{W}^{(0)} \in \mathbb{R}^{d \times H}$ is an input-to-hidden weight matrix for a hidden layer with H features, $ReLU$ () is the rectified linear activation function, and $\mathbf{W}^{(1)} \in \mathbb{R}^{H \times M}$ is a hidden-to-output weight matrix with M feature maps. M denotes the number of classes. Softmax denotes the softmax activation function. $\mathbf{E} \in \mathbb{R}^{N \times M}$ is the matrix of labels.

It motivates the choice of convolutional architecture via a localized first-order approximation of spectral graph convolutions. The GCN is used for semi-supervised classification on graph-structured data, such as citation networks or on a knowledge graph dataset.

3   Proposed Method

The GCN which is introduced in last section motivates us to jointly use the data and their associated graph in order to derive a non-linear embedding of the data and not only their label as it is the case [8]. The basic idea is to replace data samples by their convolution with a certain graph. To this end, we may use a first order approximation of spectral graph convolutions to replace the original samples data in regression model. Our goal is to estimate a final non-linear embedding in which the regressor is based on graph convolution.

Spectral convolutions on graph defined as the multiplication of a signal $\mathbf{x} \in \mathbb{R}^N$ with a filter $\mathbf{g}_w = diag(\mathbf{w})$ where $\mathbf{w} \in \mathbb{R}^N$ is parameterized in Fourier domain:

$$\mathbf{g}_w \star \mathbf{x} = U \cdot \mathbf{g}_w \cdot U^T \cdot \mathbf{x} \tag{4}$$

where $U^T\mathbf{x}$ and $\mathbf{g}_w$ could be regarded respectively as the graph Fourier transform of $\mathbf{x}$ and a function of $\mathbf{\Lambda}$ which is $\mathbf{g}(\mathbf{\Lambda})$.

The computational complexity of (4) is $\mathcal{O}(N^2)$. For solving the problem of expensive computing cost, [6] proposed Chebyshev polynomials to unfold $\mathbf{g}_w$:

$$\mathbf{g}_{w'} \approx \sum_{k=0}^{K} \mathbf{w_k}' \cdot T_k(\tilde{\mathbf{\Lambda}}) \tag{5}$$

where T_k refer to Chebyshev polynomials. $T_k(x) = 2xT_{k-1}(x) - T_{k-2}(x)$, and $T_0(x) = 1$, $T_1(x) = x$. The largest eigenvalue of the Laplacian matrix $\mathbf{L}$ is denoted by $\lambda_{\max}$. $\tilde{\mathbf{\Lambda}} = \frac{2}{\lambda_{\max}}\mathbf{\Lambda} - \mathbf{I}$. $\mathbf{w_k}'$ is Chebyshev coefficients. So the spectral convolutions on graphs with a truncated expansion in terms of Chebyshev polynomials could be rewritten:

$$\mathbf{g_{w'}} \star \mathbf{x} \approx \sum_{k=0}^{K} \mathbf{w}_k' \cdot T_k(\tilde{\mathbf{L}}) \cdot \mathbf{x} \tag{6}$$

where $\tilde{\mathbf{L}} = \frac{2}{\lambda_{\max}}\mathbf{L} - \mathbf{I}$. If we just expand 1_{st}-order polynomial in (6) to limit convolution operation, and according to further approximate in the linear formulation where $k = 1$ and $\lambda_{\max} \approx 2$ in the (6), we get:

$$\mathbf{g}_{w'} \star \mathbf{x} \approx \mathbf{w}'_0\mathbf{x} + \mathbf{w}'_1(\frac{2}{\lambda_{\max}}\mathbf{L} - \mathbf{I})\mathbf{x} = \mathbf{w}'_0\mathbf{x} - \mathbf{w}'_1\mathbf{D}^{-\frac{1}{2}}\mathbf{S}\mathbf{D}^{-\frac{1}{2}}\mathbf{x}$$
$$\Rightarrow \mathbf{w}(\mathbf{I} + \mathbf{D}^{-\frac{1}{2}}\mathbf{S}\mathbf{D}^{-\frac{1}{2}})\mathbf{x} \tag{7}$$

where $\mathbf{w}_0'$ and $\mathbf{w}_1'$ are free parameters. If we consider to constrain the number of parameters to address over-fitting and to minimize the number of matrix multiplications, $\mathbf{w} = \mathbf{w}_0' = -\mathbf{w}_1'$ could be used in the above formula (7).

We approximate $\lambda_{\max} \approx 2$ which means that the eigenvalues of $\mathbf{I} + \mathbf{D}^{-\frac{1}{2}}\mathbf{S}\mathbf{D}^{-\frac{1}{2}}$ are between 0 and 2. In fact, this operator will lead to numerical unstable. So the renormalization trick is introduced which replaces $\mathbf{I} + \mathbf{D}^{-\frac{1}{2}}\mathbf{S}\mathbf{D}^{-\frac{1}{2}}$ by $\hat{\mathbf{D}}^{-\frac{1}{2}}\hat{\mathbf{S}}\hat{\mathbf{D}}^{-\frac{1}{2}}$, where $\hat{\mathbf{S}} = \mathbf{S} + \mathbf{I}$ and $\tilde{\mathbf{D}}_{ii} = \sum_j \tilde{\mathbf{S}}_{ij}$.

In the end, we acquire the samples data $\mathbf{X} \in \mathbb{R}^{d \times N}$ and its non-linear projection data by imposing that $\mathbf{Z} = \hat{\mathbf{D}}^{-\frac{1}{2}}\hat{\mathbf{S}}\hat{\mathbf{D}}^{-\frac{1}{2}}\mathbf{X}^T \cdot \mathbf{W} \in \mathbb{R}^{N \times d}$. $\mathbf{W}$ is filter parameters matrix or transform matrix and $\mathbf{Z}$ is convolved signal matrix or regarded as regarded as projection matrix in graph theory. We also could add a bias term $\mathbf{b}$:

$$\mathbf{Z} = \hat{\mathbf{D}}^{-\frac{1}{2}}\hat{\mathbf{S}}\hat{\mathbf{D}}^{-\frac{1}{2}}\mathbf{X}^T \cdot \mathbf{W} + \mathbf{1}\mathbf{b}^T \tag{8}$$

Our proposed algorithm not only inherits the latent advantages of spectral graph convolutions which such a model can alleviate the problem of over-fitting on local neighborhood structures for graphs with wide node degree distributions, (e.g., scene and face images), but also provides the advantages of margin-based discriminant embedding and manifold smoothness. So the objective function focuses on the joint estimation of the non-linear projection data $\mathbf{Z}$, the linear transform matrix $\mathbf{W}$ and the shift vector $\mathbf{b}$. Thus, the following is minimized under a constraint:

$$h(\mathbf{Z}, \mathbf{W}, \mathbf{b}) = tr(\mathbf{Z}^T(\mathbf{L} + \lambda\tilde{\mathbf{M}}_l)\mathbf{Z}) + \mu(\|\mathbf{W}\|_2^2$$
$$+ \gamma \left\|\hat{\mathbf{D}}^{-\frac{1}{2}}\hat{\mathbf{S}}\hat{\mathbf{D}}^{-\frac{1}{2}}\mathbf{X}^T \cdot \mathbf{W} + \mathbf{1}\mathbf{b}^T - \mathbf{Z}\right\|_2^2) \quad s.t. \ \mathbf{Z}^T\hat{\mathbf{D}}_l\mathbf{Z} = \mathbf{I} \tag{9}$$

In the sequel, $\mathbf{L} + \lambda \tilde{\mathbf{M}}_l$ will be denoted by $\mathbf{V}$. λ, μ and γ are regularization parameters. The above criterion simultaneously attempts to provide graph smoothness criterion in locality preserving, margin maximization of labeled data samples, linear transform regularization ($\|\mathbf{W}\|^2$) and spectral graph convolution based regression. The final goal is to estimate a non-linear embedding and its regression transform.

In the sequel, we show how the optimal solution for function (9) can be derived.

We vanish the derivatives of the objective function (9) with respect to $\mathbf{W}$ and $\mathbf{b}$. This yields:

$$\mathbf{b} = \frac{1}{N}(\mathbf{Z}^T \mathbf{1} - \mathbf{W}^T \hat{\mathbf{X}}^T \mathbf{1}) \tag{10}$$

$$\mathbf{W} = \gamma(\gamma \hat{\mathbf{X}}^T \mathbf{H}_c \cdot \mathbf{H}_c^T \hat{\mathbf{X}} + \mathbf{I})\hat{\mathbf{X}}^T \mathbf{H}_c \cdot \mathbf{Z} \tag{11}$$

where $\hat{\mathbf{X}} = \hat{\mathbf{D}}^{-\frac{1}{2}}\hat{\mathbf{S}}\hat{\mathbf{D}}^{-\frac{1}{2}}\mathbf{X}^T$ and $\mathbf{H}_c = \mathbf{I} - \frac{1}{N}\mathbf{1}\mathbf{1}^T$. Besides, let $\mathbf{P}$ denotes the matrix $\gamma(\gamma \hat{\mathbf{X}}^T \mathbf{H}_c \cdot \mathbf{H}_c^T \hat{\mathbf{X}} + \mathbf{I})\hat{\mathbf{X}}^T \mathbf{H}_c$, thus Eq. (11) can be rewritten as $\mathbf{W} = \mathbf{P} \cdot \mathbf{Z}$. So the spectral graph regression could be deduced by:

$$\hat{\mathbf{X}}\mathbf{W} + \mathbf{1}\mathbf{b}^T = \hat{\mathbf{X}}\mathbf{P}\mathbf{Z} + \frac{1}{N}\mathbf{1}\mathbf{1}^T\mathbf{Z} - \frac{1}{N}\mathbf{1}\mathbf{1}^T\hat{\mathbf{X}}\mathbf{P}\mathbf{Z}$$

$$= (\mathbf{I} - \frac{1}{N}\mathbf{1}\mathbf{1}^T)\hat{\mathbf{X}}\mathbf{P}\mathbf{Z} + \frac{1}{N}\mathbf{1}\mathbf{1}^T\mathbf{Z}$$

$$= \mathbf{H}_c\hat{\mathbf{X}}\mathbf{P}\mathbf{Z} + \frac{1}{N}\mathbf{1}\mathbf{1}^T\mathbf{Z} \tag{12}$$

Let $\mathbf{Q} = (\mathbf{I} - \frac{1}{N}\mathbf{1}\mathbf{1}^T)\hat{\mathbf{X}}\mathbf{P} + \frac{1}{N}\mathbf{1}\mathbf{1}^T$, thus Eq. (12) can be written in a more compact form $\hat{\mathbf{X}}\mathbf{W} + \mathbf{1}\mathbf{b}^T = \mathbf{Q} \cdot \mathbf{Z}$. By plugging Eqs. (10), (11) and (12), into (9), this one becomes:

$$h = tr(\mathbf{Z}^T \mathbf{V}\mathbf{Z}) + \mu \, tr(\mathbf{Z}^T \mathbf{P}^T \mathbf{P}\mathbf{Z})$$

$$+ \mu\gamma \cdot tr((\mathbf{Q}\mathbf{Z} - \mathbf{Z})^T(\mathbf{Q}\mathbf{Z} - \mathbf{Z}))$$

$$= tr(\mathbf{Z}^T(\mathbf{V} + \mu\mathbf{P}^T\mathbf{P} + \mu\gamma(\mathbf{Q} - \mathbf{I})^T(\mathbf{Q} - \mathbf{I}))\mathbf{Z}) \tag{13}$$

The constrained optimization problem becomes:

$$\mathbf{Z} = \arg\min_{\mathbf{Z}} tr(\mathbf{Z}^T(\mathbf{V} + \mu\mathbf{P}^T\mathbf{P} + \mu\gamma(\mathbf{Q} - \mathbf{I})^T(\mathbf{Q} - \mathbf{I}))\mathbf{Z}) \; s.t. \; \mathbf{Z}^T\hat{\mathbf{D}}_l\mathbf{Z} = \mathbf{I} \tag{14}$$

Finally, the optimization problem in (14) is solved by Eigen-decomposition of $(\mathbf{V} + \mu\mathbf{P}^T\mathbf{P} + \mu\gamma(\mathbf{Q} - \mathbf{I})^T(\mathbf{Q} - \mathbf{I}))$ matrix which picks up the eigenvectors corresponding to the smallest eigenvalues.

Once the non-linear projection $\mathbf{Z}$ is estimated, $\mathbf{W}$ and $\mathbf{b}$ could be obtained using Eqs. (10), (11). Algorithm 1 summarizes the main steps of the proposed

---

Algorithm 1. Proposed Algorithm

Input: Data matrix: $\mathbf{X}$; parameters: μ, λ and γ.
Output: Non-linear embedding matrix $\mathbf{Z}$; linear transform $\mathbf{W}$ and bias term $\mathbf{b}$;
1: Estimate the sparse graph matrix $\mathbf{S}$ using (1);
2: Renormalize the matrix $\hat{\mathbf{S}} = \mathbf{I} + \mathbf{S}$ using $\hat{\mathbf{D}}^{-\frac{1}{2}}\hat{\mathbf{S}}\hat{\mathbf{D}}^{-\frac{1}{2}}$;
3: Compute the spectral graph convolutions matrix $\hat{\mathbf{X}} = \hat{\mathbf{D}}^{-\frac{1}{2}}\hat{\mathbf{S}}\hat{\mathbf{D}}^{-\frac{1}{2}}\mathbf{X}^T$;
4: Update $\mathbf{Z}$, $\mathbf{W}$ and $\mathbf{b}$ using Eqs. (14), (11) and (10) respectively;
5: Compute the non-linear embedding matrix $\mathbf{Z}$ as $\mathbf{Z} = \hat{\mathbf{X}} \cdot \mathbf{W} + \mathbf{1}\mathbf{b}^T$;

---

algorithm. If an unseen test data $\mathbf{x}_u \in \mathbb{R}^{d \times 1}$ is given, the sample data graph-embedding is obtained by $\mathbf{z}_u = \mathbf{W}^T\hat{\mathbf{x}}_u + \mathbf{b}$. $\hat{\mathbf{x}}_u$ is computed as follows. First, we compute the edge weights between $\mathbf{x}_u$ and the original data samples $\mathbf{X}$. This allows to expand the original graph matrix, $\mathbf{S}$, by one row and one column. This expanded graph matrix is then renormalized in the same way described above. $\hat{\mathbf{x}}_u$ is obtained by $A_{N+1} X'^T$ where A_{N+1} denotes the last row in the normalized graph matrix and X'^T is the augmented data matrix.

It is worth noting that although our model is inspired from the concept of Graph Convolution [8], it has several differences with the latter one. First, our work addresses flexible non-linear data embedding while the work of [8] addresses label propagation. Second, the application domain in [8] is the semi-supervised document classification for which binary graphs are defined. In our work, we address image datasets for which similarity graphs are more challenging to estimate.

4   Experimental Results

This section presents the performance of the proposed embedding on different kinds of image data which include scene, face and object datasets. The datasets used are 8 Sports Event Categories dataset, Scene 15 dataset, Extended YALE Face dataset and COIL-20 object dataset. In the 8 Sports Event Categories and Scene 15 datasets, we use block-based Local Binary Patterns [14] as image descriptor. For these two datasets, the number of blocks is set to 10×10 and the LBP descriptor is the uniform one having 59 features. For the face datasets, due to the small size of the face images, we use image raw brightness as image descriptor.

4.1   Datasets

8 Sports Event Categories Dataset: The 8 sports event categories dataset [11] is provided by Li and Fei-Fei. This dataset contains 8 sports event categories: rowing (250 images), badminton (200 images), polo (182 images), bocce (137 images), snowboarding (190 images), croquet (236 images), sailing (190 images), and rock climbing (194 images). These images are high-resolution. We use 130 images in every category, The total number of images is 1040 images.

We randomly select 50% and 70% of data as the training set and use the remaining 50% and 30% of data as the test set.

Scene 15 Dataset: Scene 15 dataset [10] contains 4485 gray images of 15 different scenes including both indoor scenes and outdoor scenes. The dataset does not provide separated training and test sets. We use 130 images in every category. The total number of images is 1950 images. We randomly select 50% and 70% of data as the training set and use the remaining 50% and 30% of data as the test set.

Extended YALE Face Dataset: The cropped version that contains 38 individuals has been used in our experiments. The images of the cropped version contain illumination variation and facial expression variation. The images are in gray scale, and we have rescaled them to 32×32 pixels in our experiments. We use a subset of the database containing 50 images for each person, and randomly select 20% and 40% of data as the training set and use the remaining 80% and 60% of data as the test set.

COIL-20 Object Dataset: This dataset (Columbia Object Image Library) [13] consists of 1440 images of 20 objects. Each object has 72 images corresponding to 72 rotations of the object, where the rotation angle between two images is 5 degrees. We use a subset of images containing 70 images for each object, and randomly select 10% and 20% of data as the training set and use the remaining 90% and 80% of data as the test data.

4.2   Experimental Setup

We compare the proposed framework with several state-of-the-art algorithms: Flexible Semi-Supervised Embedding algorithm (FSSE) [4], Manifold Regularized Deep Learning Architecture Algorithm (MRDL) [19], Kernel Flexible Model Embedding (KFME) [5], Joint Embedding Learning and Sparse Regression (JELSR) [7], Supervised Laplacian Eigenmaps (SLE) [15], Semi-Supervised Discriminant Analysis (SDA) [2], Semi-Supervised Discriminant Embedding (SDE) [18], LLE [16]. All the above methods provide data embedding except the KFME method which is a label propagation method. Once the embedding is computed data are classified in the obtained space using the Nearest Neighbor Classifier (NN).

In Sect. 2.2, we have briefly introduced the MRDL algorithm. It is used with two layers in our tests. Besides, Sect. 2.2 proposed a kernel sparse algorithm for acquiring similarity matrix $\mathbf{S}$ which is also used in our paper. We use the normalized graph Laplacian matrix $\mathbf{L} = \mathbf{I} - \mathbf{D}^{-\frac{1}{2}}\mathbf{S}\mathbf{D}^{-\frac{1}{2}}$. For the JELSR and FSSE methods, we use two types of graphs: the classic KNN graph and the kernel sparse graph of [19] to compare the performance of the different graphs. The proposed method has three balance parameters: λ, μ and γ. We set each parameter to a subset of values belonging to $\{10^{-3}, 10^{-2}, 1, 10^{-1}, 1, 10^1, 10^2, 10^3, 10^6, 10^9\}$. We then report the top-1 recognition accuracy (best average recognition rate) of all methods from the best parameter configuration. All results are obtained with ten random splits of the data into a train set and a test set. For train sets, two different percentages are considered for every dataset.

According to [7], JELSR algorithm is used for unsupervised feature selection by computing the scores of all features. It also can be used for graph-based embedding.

4.3   Method Comparison

The performance of the different competing methods are summarized in Table 1. The results correspond to 4 different datasets and different train data percentages. These results correspond to an average over 10 random splits.

Figure 1 depict the average performance of the competing methods (KFME, LLE, JELSR, MRDL, SSFE and proposed algorithm (FDEFS)) as a function of the number of non-linear features. These figures respectively correspond to the 8 Sports and Scene 15 datasets. The KFME algorithm does not depend on the feature dimension since it is a label propagation method. Indeed, the maximum dimension of SDA method is given by $C - 1$, where C is the number of classes.

Table 1. Best average recognition rate (%) obtained on the 8 Sports Event and Scene 15 dataset using 10 random splits with two different percentages for the training part.

Dataset	8 Sports event		Scene 15 dataset		Extended Yale B		COIL-20	
Method	P = 50%	P = 70%	P = 50%	P = 70%	P = 20%	P = 40%	P = 10%	P = 20%
LLE	54.92	59.10	44.26	47.42	91.47	95.75	91.81	94.71
SLE	51.40	50.90	50.48	50.65	83.20	93.39	82.03	88.56
SDA	63.46	66.06	61.52	63.73	89.96	96.54	95.33	98.07
SDE	51.98	55.96	46.10	48.07	85.92	92.76	89.10	95.33
MRDL	51.77	52.85	46.59	47.91	76.78	78.97	88.00	88.86
KFME	62.58	65.03	60.89	63.74	90.39	92.85	**96.98**	98.56
JELSR	55.92	57.60	51.83	58.59	85.31	90.13	93.80	96.88
JELSR (KNN)	52.46	55.48	41.37	44.24	75.03	84.54	85.48	93.01
FSSE	64.72	67.18	64.78	68.17	93.71	**98.31**	93.60	97.83
FSSE (KNN)	50.96	63.24	50.96	55.62	93.36	98.18	86.19	93.61
Proposed algorithm	**67.04**	**69.97**	**67.26**	**70.36**	**93.88**	97.53	95.47	**98.76**

From the obtained results depicted in the previous tables and figures, we can draw the following conclusions:

- Our proposed algorithm has achieved very good performance in all four datasets: 8 Sports Event Categories Dataset, Scene 15 Dataset, Extended YALE Face Dataset and COIL-20 Object Dataset. For the first two dataset the superiority of the proposed method is obvious. This can be explained by the fact the use of Graph Convolution principle in the regression function can tackle the class high variability.

The proposed method was slightly outperformed by KFME on COIL-20 dataset corresponding to the 10% train data experiment and FSSE on Extended YALE dataset corresponding to the 40% train data experiment.

- By inspecting the results obtain by FSSE, FSSE (KNN), JELSR, JELSR (KNN), we can see that the kernel sparse graph used in graph smoothness has significantly improved the performance of the same frameworks that use the classic KNN graph.
- From Fig. 1, we can observe that by increasing the feature dimension the rate cannot be improved. In fact, the highest rate always happen with very few features indicating that the proposed method has achieved a very good dimensionality reduction.

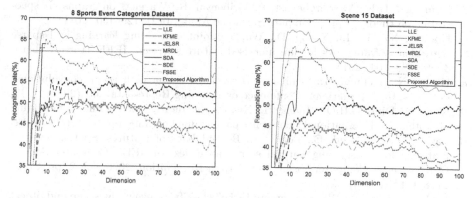

Fig. 1. Recognition accuracy vs. feature dimension for 8 Sports Event Categories Dataset and Scene 15 Dataset. Test samples per class were 50%. The classifier used was 1-NN.

5   Conclusion

We proposed a framework for discriminative non-linear Graph-based embedding with Spectral Graph Convolutions Structure. This framework can solve the over-fitting on local neighborhood structures for graphs. The framework combine with many criteria: manifold smoothness, Margin Discriminant Embedding and regression with Graph Spectral Convolutions. Experiments on scene, face and object image datasets have shown the outstanding ability of the model with superiority with respect to many competing algorithms.

Future work will investigate two directions. On the one hand, we will consider other varieties of regression, even sparse regression. On the other hand, we will explore more effective Graph Spectral Filter or polynomial expansion algorithm. Moreover, the algorithm could be used for more types of data, such as social networks, citation networks, knowledge graphs and many other real world graph datasets.

References

1. Belkin, M., Niyogi, P.: Laplacian eigenmaps and spectral techniques for embedding and clustering. In: Advances in Neural Information Processing Systems, pp. 585–591 (2002)
2. Cai, D., He, X., Han, J.: Semi-supervised discriminant analysis. In: 2007 IEEE 11th International Conference on Computer Vision, pp. 1–7 (2007)
3. Chen, H.T., Chang, H.W., Liu, T.L.: Local discriminant embedding and its variants. In: IEEE Computer Society Conference on Computer Vision and Pattern Recognition, CVPR 2005, vol. 2, pp. 846–853. IEEE (2005)
4. Dornaika, F., El Traboulsi, Y.: Learning flexible graph-based semi-supervised embedding. IEEE Trans. Cybern. 46(1), 206–218 (2016)
5. El Traboulsi, Y., Dornaika, F., Assoum, A.: Kernel flexible manifold embedding for pattern classification. Neurocomputing 167, 517–527 (2015)
6. Hammond, D.K., Vandergheynst, P., Gribonval, R.: Wavelets on graphs via spectral graph theory. Appl. Comput. Harmonic Anal. 30(2), 129–150 (2011)
7. Hou, C., Nie, F., Li, X., Yi, D., Wu, Y.: Joint embedding learning and sparse regression: a framework for unsupervised feature selection. IEEE Trans. Cybern. 44(6), 793–804 (2014)
8. Kipf, T.N., Welling, M.: Semi-supervised classification with graph convolutional networks. In: International Conference on Learning Representations (2017)
9. Kong, D., Ding, C.H.Q., Huang, H., Nie, F.: An iterative locally linear embedding algorithm. In: International Conference on Machine Learning, pp. 1647–1654 (2012)
10. Lazebnik, S., Schmid, C., Ponce, J.: Beyond bags of features: Spatial pyramid matching for recognizing natural scene categories. In: IEEE Computer Society Conference on Computer Vision and Pattern Recognition, 2006, vol. 2, pp. 2169–2178. IEEE (2006)
11. Li, L.J., Fei-Fei, L.: What, where and who? classifying events by scene and object recognition. In: IEEE 11th International Conference on Computer Vision, ICCV 2007, pp. 1–8. IEEE (2007)
12. Martínez, A.M., Kak, A.C.: pCA versus LDA. IEEE Trans. Pattern Anal. Mach. Intell. 23(2), 228–233 (2001)
13. Nene, S.A., Nayar, S., Murase, H.: Columbia object image library (coil-20). Technical report CUCS-005-96 (1996)
14. Ojala, T., Pietikäinen, M., Harwood, D.: A comparative study of texture measures with classification based on featured distributions. Pattern Recogn. 29(1), 51–59 (1996)
15. Raducanu, B., Dornaika, F.: A supervised non-linear dimensionality reduction approach for manifold learning. Pattern Recogn. 45(6), 2432–2444 (2012)
16. Roweis, S.T., Saul, L.K.: Nonlinear dimensionality reduction by locally linear embedding. Science 290(5500), 2323–2326 (2000)
17. Tenenbaum, J.B., De Silva, V., Langford, J.C.: A global geometric framework for nonlinear dimensionality reduction. Science 290(5500), 2319–2323 (2000)
18. Yu, G., Zhang, G., Domeniconi, C., Yu, Z., You, J.: Semi-supervised classification based on random subspace dimensionality reduction. Pattern Recogn. 45(3), 1119–1135 (2012)
19. Yuan, Y., Mou, L., Lu, X.: Scene recognition by manifold regularized deep learning architecture. IEEE Trans. Neural Netw. Learn. Syst. 26(10), 2222–2233 (2015)

Graph-Based Representations for Supporting Genome Data Analysis and Visualization: Opportunities and Challenges

Vincenzo Carletti[1], Pasquale Foggia[1(✉)], Erik Garrison[2], Luca Greco[1],
Pierluigi Ritrovato[1], and Mario Vento[1]

[1] DIEM, University of Salerno, Fisciano, Italy
{vcarletti,pfoggia,lgreco,pritrovato,mvento}@unisa.it
[2] Genomics Institute, University of California Santa Cruz, Santa Cruz, USA
erik.garrison@gmail.com

Abstract. Genetics has known an extraordinary development in the last years, with a reduction of several orders of magnitude in the costs and the times required to obtain the sequence of nucleotides corresponding to a whole genome, leading to the availability of huge amounts of genomic data. While these data are essentially very long strings, several graph-based representations have been introduced to perform efficiently some operations on a single genome or on a set of related genomes. In this paper we will review the most important types of genetic graphs, together with the algorithmic challenges and open issues related to their use.

Keywords: Graph representation of genomic data · Sequence graphs · De Bruijn graphs · Genome graphs

1 Introduction

Deoxyribonucleic acid (DNA) encodes genetic instructions used in the growth, development, functioning and reproduction of any organism. DNA molecules, called *chromosomes* structured as two twisted strands of nucleotides arranged in a double helix form. Each DNA strand is composed of four kinds of nitrogen-based nucleotides: Cytosine, Guanine, Adenine and Thymine (shortened as C, G, A and T). Two separate strands are bound together according to base pairing rules: A with T and C with G. Thus, given the sequence of one strand, the complementary sequence of the other one can be automatically derived. The length of a DNA sequence is measured in *base pairs* (bp).

The era of DNA sequencing begun in 70s when F. Sanger developed a reliable method based on the DNA polymerization chain-termination concept, where a DNA strand is broken in several fragments, called *reads* (usually 300bp to 1000bp), and each of them is decoded using a sequential process. In the early 2000s Next Generation Sequencing (NGS) approaches (also known as

© Springer Nature Switzerland AG 2019
D. Conte et al. (Eds.): GbRPR 2019, LNCS 11510, pp. 237–246, 2019.
https://doi.org/10.1007/978-3-030-20081-7_23

High Throughput Sequencing) were proposed, where many short reads (50bp to 250bp) are produced and analyzed in parallel, reducing per-base costs and computational time dramatically (nowadays, a whole human genome can be sequenced in 44 h with a cost of about $1000).

Many genome analysis techniques use a data representation based on graphs: sequencing and alignment are two important examples. In this paper, after a brief introduction to the different DNA analysis problems (Sect. 2), we will describe the most important types of graphs that are used for supporting the solution of these problems (Sect. 3) and will provide a short overview of the tools adopting these graphs (Sect. 4). Then we will discuss about the challenges and open issues in Sect. 5.

2   DNA Analysis Problems

2.1   Genome Assembly

Sequencing machines (both those based on Sanger's approach and those adopting NGS) produce a large set of overlapping reads, that are subsequences of the desired chromosome sequence of bases; these reads are relatively long for Sanger's method, and much shorter (even 50bp) for NGS. NGS machines, additionally, produce reads from both strands of DNA at the same time, thus adding the complexity of attributing each read to the correct strand.

The first step to be accomplished in order to reconstruct the whole genome, is the composition of all the reads into a single sequence; this process is known as *assembly*. The assembly problem has to variants: the first, called *de novo* assembly, can only use the information obtained from the reads. This is the case when the DNA of a species is sequenced for the first time. The other variant is *resequencing*: a reference genome for the same species is available, and is used to guide the choices of the sequencing algorithm. A subtask of resequencing is the matching of a read with one or more reference genomes: this is called *mapping*.

As detailed in [2,18], the most adopted schemes for de novo assembly, and many schemes for resequencing, are based on some graph representation.

2.2   Sequence Alignment

Given two sequences, *Sequence Alignment* is the process of finding the correspondence between their positions that minimizes the number of edit operations required to eliminate the differences between the two sequences. A generalization to more than two sequences is called *Multiple Sequence Alignment* (MSA).

Exact alignment algorithms exist (e.g. Smith and Waterman's), but their complexity is $O(NM)$ (N and M are the lengths of the two sequences); thus they are not practically applicable to large genomic data.

Suboptimal solutions can be found by indexing either the query or target set of sequences to efficiently obtain patterns of exact matches; with the generated indices, candidate sub-regions can be quickly found and examined for more sensitive alignment. This intuition is at the basis of the two famous alignment algorithms FASTA and BLAST. In recent approaches, especially for MSA problems, graphs-based data structures like *De Bruijn* and *Cactus* graphs (see Sect. 3) have been successfully used [4,19].

2.3 Variant Calling and Pangenomics Analysis

Given a population of individuals belonging to the same species, their genomes have small differences. *Variant Calling* is the process of determining, for each individual, the set of variations with respect to a reference genome. Competition fostered by the *1000 Genome Project* [17] encouraged the development of variant calling algorithms based on a variety of principles.

An efficient representation of the whole population (the *pangenome*) is the starting point for more complex kinds of analysis, related to the population or to the species as a whole, that form the domain of *pangenomics* [1]. Several recent approaches advocate for the use of graphs for representing a set of genomes [5,13].

3 Graph-Based Representation for DNA Sequences

3.1 Overlap Graphs

The starting point for defining an *overlap graph* is a set of segments $S = s_1, \ldots, s_n$, where each segment contains a string of nucleotides obtained by a read. The segments in general do not have the same length.

Given two segments s and t, the *maximal proper overlap* $ov(s,t)$ is defined as the longest string y such that $s = x|y$ and $t = y|z$, where $|$ is the string concatenation operator, and x and y must be non-empty strings. An error-tolerant version can be defined for taking into account a small number of reading errors; this function can be easily computed using dynamic programming.

An *overlap graph* is a directed graph $G = (V, E)$ which has n nodes $v_1, \ldots, v_n \in V$, each labeled with a segment $l(v_i) = s_i$. Two nodes v_i and v_j are connected with an edge $e_{ij} \in E$ if and only if:

$$k = |ov(s_i, s_j)| > \tau \qquad (1)$$

where τ is a suitably defined threshold. The edge is labeled with $l(e_{ij}) = k$.

Overlap graphs have been used in one of the earliest graph-based sequencing approaches, named *Overlap-Layout-Consensus* (OLC, [6]). They have been the standard sequencing algorithm for projects adopting Sanger sequencing method, that produces a (relatively) small number of long segments (e.g. the Human Genome Project worked with 150 kb segments).

In OLC, sequencing hypotheses are generated as maximal Hamiltonian paths inside the overlap graph, i.e. paths that traverse each node at most once, usually with additional constraints derived from additional knowledge about the problem (e.g. certain edges must be present in the path). Since the Hamiltonian path problem is known to be NP-Hard, heuristics are used to find a solution in an acceptable time.

3.2   String Graphs

Overlap graphs are a suitable representation for information coming from first generation sequencers, but they have some problems with the Next Generation Sequencing approach. First, the NGS sequencers obtain samples from both strands of DNA, thus each segment must be considered equivalent to its reversed complement (e.g. the segment ATATCGA must be considered equivalent to TCGATAT, obtained reversing the segment and replacing every base with its paired one). A second issue is that the length of the segments is shorter, and this creates a problem with repeated sequences: the overlap graph cannot represent unambiguously repetitions whose length is greater than the length of the segments covering them.

The string graph, introduced by Myers in 2005 [11], is an attempt at modifying the overlap graph for taking into account these problems.

In the string graph, for each segment there are two nodes, representing the endpoints of the segment. The nodes are connected by directed edges, which are associated to sequences of nucleotides; an edge can be also traversed in the reverse direction, and in that case it represents the reversed complemented sequence. The overlaps between segments are represented by additional edges linking their endpoints.

An additional information associated to each edge is its multiplicity, which is inferred in a preprocessing step from the analysis of the frequencies of observations of each segment, compared with a probabilistical model of sequence repetitons.

In this way, the string graph can represent in an unambiguous way all the information that is obtained from a NGS sequencer. The assembly is performed by looking for a modified form of Eulerian paths in the graph (i.e. paths that traverse the edges in the graphs with constraints on the minimum and maximum number of times each edge is traversed). This is a problem that can be solved in polynomial time (in the worst case, the time is quadratic with respect to the number of edges). Furthermore, an important property of string graphs is that every path is *read coherent*, i.e. it represents a valid assembly of the obtained reads; other types of graphs (e.g. the De Bruijn graphs) lack this property, and thus require additional computational steps to filter out incoherent paths.

3.3   De Bruijn Graphs

Properly called De Bruijn graphs are a formal tool introduced in the 1940s to solve the mathematical problem of finding the shortest string containing all the possible substrings of length k over a given alphabet. Since this problem has similarities to the sequencing problem, and the solution based on De Brujn graphs is quite effective, a slightly modified version of these graphs has been introduced in the late 1990s for DNA data [14].

De Bruijn graphs, or DBG, are based on k-*mers*, which are substrings of length k of a read. Dividing the reads into k-mers is not expected to cause a significant loss of information with NGS sequencers, since they produce reads that are already short, and with a very small per-base error probability (so each k-mer is almost certainly correct). The advantage of working with k-mers is that they can be efficiently encoded (e.g. with a single 32 bit word for $k = 16$), allowing the use of fast binary operations for comparing and manipulating them.

For a given k, the nodes in a De Bruijn graph are associated to the $(k - 1)$-mers found in the reads. For each k-mer s in the reads, a directed edge (v_1, v_2) is added to the graph between the node v_1 containing the first $k - 1$ bases of s, and v_2 containing the last $k - 1$ bases of s. For example, if $k = 4$ and $s =$ ATAG, an edge is added between the node ATA and the node TAG. In some variants (colored DBG) additional information is added to the edges to represent the frequency of a k-mer.

Once the DBG has been built from a set of reads, the sequencing problem can be solved using Eulerian paths that, as mentioned in the previous subsection, can be found with fast algorithms. Unfortunately, while the correct sequencing is represented by a Eulerian path in the DBG, not every Eulerian path yields a valid sequencing; additional heuristics are used to filter out unvalid paths.

3.4   Genome Alignment Graphs

For the problem of (Multiple) Sequence Alignment described in Sect. 2.2, *alignment graphs* have been proposed as a suitable representation by [15]. An alignment graph G_a has a node for every base in each of the sequences being aligned. Two nodes belonging to different sequences are connected by an undirected edge if an alignment has been established between the corresponding bases. Also, the graph nodes have a partial ordering relation $\prec$, with $v_1 \prec v_2$ iff they are in the same sequence and v_1 precedes v_2.

A more compact representation of an alignment is provided by *base graphs*, that can be constructed from alignment graphs: a base graph G_b has a node for each connected component of G_a; thus a node v^b in G_b represents a set of nodes in G_a. A directed edge (v_1^b, v_2^b) is added to G_b if there are two G_a nodes, $n \in v_1^b$ and $m \in v_2^b$, such that $n \prec m$.

Enredo graphs are a generalization of alignment graphs to manage bidirectional alignments; a further extension of Enredo graphs are the *Cactus graphs* [12], whose nodes correspond to tree-like substructures in the base graph.

3.5   Tiled Graphs and Sequence Graphs

After the Multiple Sequence Alignment has been performed for the genomes of a set of individuals, a compact graph-based representation of the genomes of the considered population (the so-called *pangenome*) can be constructed, avoiding the repetition of the parts common to several individuals. This is useful to efficiently compare a new genome with those of the whole population (pangenome alignment) and for other genetic investigations.

The first graph-based representation of a pangenome has been the *tiled graph* [16]. In a tiled graph, the aligned genomes are broken into small, fixed-size pieces (tiles), which are associated to the nodes of the graph. A tile that is present in several genomes is only represented once in the graph. However, if the same subsequence is present at several positions in a genome, it is represented by different nodes. A directed edge (v_1, v_2) is added if in at least one of the genomes, the tile associated to v_1 is followed by the one of v_2. The resulting graph is directed and acyclic.

The main inconvenient of tile graphs is the inefficiency related to the fixed tile size: if the size is too small, a large number of nodes is required, even in long parts of the pangenome that are common to the whole population; on the other hand, whenever the genomes differ in a position (even by a single base), the whole tile has to be duplicated, and this lead to an inefficient redundancy when the tiles are large.

Sequence graphs [1] overcome the limitations of tiled graphs, by associating variable-length sequences of bases to each node. Thus, long subsequences that are common to all the members of the population are represented by a single node in the graph. On the other hand, when there is a single-base variation in a certain position (a so-called Single-Nucleotide Polymorphism), the corresponding node only contains the changed base. Directed edges represent the adjacency relation, in a similar way to tiled graphs.

The simplest form of sequence graph is the sequence DAG (Directed Acyclic Graph). In this kind of sequence graph, the repetition of subsequences in a genome (a phenomenon known as Copy Number Variants) requires the duplication of nodes. In the general (cyclic) version of the sequence graphs, these repetitions can be represented by introducing cycles in the graph (see Fig. 1 for an example).

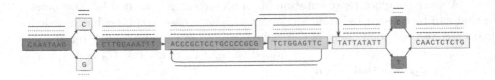

Fig. 1. An example of a cyclic sequence graph.

3.6   Variation Graphs

Variation graphs [5] have been recently introduced as a representation for pangenomes, and can be considered as an extension of sequence graphs.

A variation graph $G = (V, E, P)$ is composed of three entities:

- a set V of nodes; a node v is associated to a sequence $seq(v)$ of bases; however a node can be traversed also in the reverse direction, and in this case it represents the complemented reversed sequence; we will denote as $\overline{v}$ the node v in the reverse direction;

- a set E of directed edges; also edges can be traversed in the reverse direction, and the reversal of (v_1, v_2) is equivalent to edge $(\overline{v_2}, \overline{v_1})$; a variation graph can contain reversing edges (edges like $(v_1, \overline{v_2})$), self-loops (edges like (v, v)) and reversing loops (edges like $(v, \overline{v})$);
- a set P of paths; each path in P correspond to one of the genomes used to build the graph; the addition of this information makes possible the precise reconstruction of the original genomes, which otherwise would not be possible. Paths are also used in the implementation of edit operations on variation graphs.

A variation graph retains all the information present in the original collection of genomes, and can represent in a compact way all the kinds of variations (for instance, an individual may have a copy of a subsequence that is reversed with respect to the other individuals).

4 Graph Based Tools for Genome Analysis

In this section we overview the most relevant tools leveraging an internal graph representation for addressing genome read mapping and processing.

In the context of mapping, GenomeMapper [16] is one of the first effective tools supporting simultaneous mapping of short reads against multiple genomes by integrating several genomes into a single graph structure. Besides the efficiency improvement, authors have demonstrated also a reduction in the reference bias. Limasset et al. [7] have proposed an approach relying on a pipeline called GGMAP (Greedy Graph MAPping). The novelty of the contribution is the procedure to map reads on branching paths of the graph, that exploits a custom designed heuristic algorithm called BGREAT (de Bruijn Graph REAd mapping Tool). The GGMAP approach has shown the ability to map millions of reads per CPU hour on a De Bruijn graph built from a large set of human genomic reads. In [8] Liu et al. have introduced the *de Bruijn Graph-based Aligner* (deBGA), a graph-based seed-and-extension algorithm to align NGS reads to a reference genome, organized and indexed using a De Bruijn graph. deBGA has proven to obtain good sensitivity and accuracy without penalizing speed, thanks to the advanced handling of repeats.

Maciuca et al. [9] propose *Gramtools*, using a form of Sequence DAGs for representing a pangenome, with the addition of indexing structures based on the *FM-index*. This approach is simple and allows for a linear-time indexing of the pangenome, but it still presents a query time that is 100 times slower than competing methods. Eggertson et al. have proposed the tool *GraphTyper* [3], that uses a standard short read aligner to align Illumina reads against the reference genome. Whereas alignments present soft clips and apparent differences from the reference, they are matched to a sequence DAG built from the reference and variant calling data. GraphTyper uses the graph reference system internally to improve algorithm performance, but the results are projected back into the linear reference. The method has efficient and accurate performance on exceptionally large resequencing problems, justifying its utility of the graph realignment approach.

In recent times, there is an increasing trend in using haplotype information to retrieve the constrained sequence search space of a graph-based reference genome. This is the case of CHOP [10], a path indexer for graphs that incorporates haplotype-level information to constrain the resulting index. It decomposes the graph into a set of linear sequences, so that reads can be aligned by established methods, such as BWA or Bowtie2. The approach enables typical sequence aligners to perform read alignments to graphs that store any type of variation. The method can achieve increased sensitivity for variation detection by iteratively integrating variation into a graph and can be applied to large and complex datasets (the authors apply it on a graph-based representation of chromosome 6 of the human genome encoding the variants reported by the 1000 Genomes Project). CHOP, on the other hand, can only use short-range haplotype information in read mapping.

5   Main Issues and Challenges for Use of Graph for DNA Analysis

The open problems related to graphs in genomics are essentially of three kinds: first, finding an efficient representation for the graphs; second, improve the algorithm performance, both in terms of speed and in terms of accuracy; third, develop new operations that work on the graphs as a whole.

Regarding the first problem, consider that the graphs used in genomics have typically at least hundreds of millions nodes. The data coming out of a sequencer for a single human genome are around 200 Gb, and if the graph must preserve all the information in the reads (as is required for some problems, such as de novo alignment), you have to keep this information, even if it is redundant (the same part of the sequence is covered by several reads); furthermore you have to add the edges to the graph, and usually you need some kind of index data structure to quickly find the nodes containing a given substrings. Since it is important to ensure that the whole graph can fit in RAM (not on a typical workstation; you need to have a well endowed cluster, with, say 256 Gb of memory), a significant reasearch effort is being directed at techniques of graph compression that reduce the size of the graph without loosing information (or too much of it).

The second problem is related to the fact that the main graph-related tasks (mapping and alignment) are performed using heuristic algorithms with no guarantee of optimality. For instance, mapping uses index structures to quickly search a string of bases within the nodes of the graph; however, these indices often have problems when the string is present but split between two adjacent nodes. This may cause a false miss that would not have happened if the string were matched against a linear reference genome. Also, exact matching is a problem when the read is affected by errors (which are more common in NGS sequencers). The challenge here is both to provide new heuristic algorithms or index data structures that can improve the precision of the mapping, and to make these algorithms faster, so as to be applicable to larger graphs (e.g. pangenomes derived from larger populations). Regarding the speed, some of the existing tools are able to use a limited parallelism, assuming a shared memory architecture. One challenging question is if they can be modified to support a massively distributed processing model, like the Map-Reduce paradigm made popular by Google.

The third problem is related to the fact that many operations still rely on the conversion of graph data into another format so as to be able to use established algorithms, because of the lack of corresponding algorithms defined in the space of the graphs. For example, a genome graph can be constructed from a set of genomes; however, if we have two genome graphs, with the actual tools the only way to join them so as to represent the union of the two populations is to reconstruct the original sequences from the two genome graphs and then re-run the graph building algorithm from scratch. Also, while there are tools for comparing an individual with a population represented by a genome graph, there are not yet established tools for comparing two populations. Two examples of graph operations that have been successfully applied in other fields are the subgraph isomorphism and the computation of graph edit distance. While for both these operations there are heuristic algorithms, none of them appear currently to be able to support such large graphs. If is it possible to devise such algorithms, is another open question for the research community.

6 Conclusions

DNA data are becoming increasingly easy to obtain in large amounts, and their processing is important for several emerging applications in medicine and biology. While DNA data are essentially strings, non-linear, graph-based representations have been used for several years for problems like the sequencing and the alignment, and are becoming very important for applications where populations (as opposed to single individuals) have to be considered.

We have reviewed the more important kinds of graphs, highlighting how and why they are used in genomic research, and discussing the open issues and challenges for the future. More compact representations, faster and more accurate algorithms, but especially the possibility of operating on the graph structures as a whole, may contribute to build the future fundamental tools of new research fields like pangenomics.

References

1. Computational pan-genomics: status, promises and challenges. Briefings Bioinform. **19**(1), 118–135 (2016)
2. Blazewicz, J., et al.: Graph algorithms for DNA sequencing-origins, current models and the future. Eur. J. Oper. Res. **264**(3), 799–812 (2018)
3. Eggertsson, H.P., et al.: Graphtyper enables population-scale genotyping using pangenome graphs. Nat. Genet. **49**(11), 1654–1660 (2017)
4. Fostier, J., et al.: A greedy, graph-based algorithm for the alignment of multiple homologous gene lists. Bioinformatics **27**(6), 749–756 (2011). https://doi.org/10.1093/bioinformatics/btr008

5. Garrison, E., et al.: Sequence variation aware genome references and read mapping with the variation graph toolkit. bioRxiv, p. 234856 (2017)
6. Kececioglu, J.D., Myers, E.W.: Combinatorial algorithms for DNA sequence assembly. Algorithmica **13**, 7–51 (1995)
7. Limasset, A., Cazaux, B., Rivals, E., Peterlongo, P.: Read mapping on DeBruijn graphs. BMC Bioinform. **17**(1), 237 (2016). https://doi.org/10.1186/s12859-016-1103-9
8. Liu, B., Guo, H., Brudno, M., Wang, Y.: Debga: read alignment with De Bruijn graph-based seed and extension. Bioinformatics **32**(21), 3224–3232 (2016)
9. Maciuca, S., del Ojo Elias, C., McVean, G., Iqbal, Z.: A natural encoding of genetic variation in a burrows-wheeler transform to enable mapping and genome inference. In: Frith, M., Storm Pedersen, C.N. (eds.) WABI 2016. LNCS, vol. 9838, pp. 222–233. Springer, Cham (2016). https://doi.org/10.1007/978-3-319-43681-4_18
10. Mokveld, T.O., Linthorst, J., Al-Ars, Z., Reinders, M.: Chop: haplotype-aware path indexing in population graphs. bioRxiv p. 305268 (2018)
11. Myers, E.W.: The fragment assembly string graph. Bioinformatics **21**(suppl. 2), ii79–ii85 (2005)
12. Paten, B., et al.: Cactus graphs for genome comparisons. J. Comput. Biol. **18**(3), 468–481 (2011)
13. Paten, B., Novak, A.M., Eizenga, J.M., Garrison, E.: Genome graphs and the evolution of genome inference. Genome Res. **27**(5), 665–676 (2017)
14. Pevzner, P.A., Tang, H., Waterman, M.S.: An Eulerian path approach to DNA fragment assembly. In: Proceedings of the National Academy of Sciences, vol. 98, pp. 9748–9753 (2001)
15. Rausch, T., Emde, A.K., Weese, D., Döring, A., Notredame, C., Reinert, K.: Segment-based multiple sequence alignment. Bioinformatics **24**(16), i187–i192 (2008)
16. Schneeberger, K., et al.: Simultaneous alignment of short reads against multiple genomes. Genome Biol. **10**(9), R98 (2009)
17. Siva, N.: 1000 genomes project (2008)
18. Wajid, B., Serpedin, E.: Review of general algorithmic features for genome assemblers for next generation sequencers. Genomics, Proteomics Bioinform. **10**(2), 58–73 (2012)
19. Zerbino, D., Birney, E.: Velvet: algorithms for de novo short read assembly using De Bruijn graphs. Genome Res. **18**(5), 821–829 (2008). gr-074492

Author Index

Printed in the United States
By Bookmasters